The Simon and Schuster Handbook of

Anatomy and Physiology

Dr. James Bevan

Simon and Schuster, New York

Contents

Advisory Panel

Dr J. H. Baron, DM, FRCP
Senior Lecturer and Consultant, Royal Postgraduate
Medical School and Hammersmith Hospital, London

Professor G. M. Besser, MD, FRCP
Professor of Endocrinology, St Bartholomew's Hospital, London

Mr Michael Cameron, FRCS, FRCOG
Consultant Obstetrician and Gynecologist,
St Thomas' Hospital, London

Dr Desmond Croft, DM, FRCP
Consultant Physician, St Thomas' Hospital, London

Mr D. Garfield Davies, FRCS
Consultant Otorhinolaryngologist, Middlesex Hospital, London

Dr Max Friedman, MD, PhD, FRCP
Consultant Pediatrician,
University College Hospital, London

Mark Oberg-Brown, PhD

Shirley Orbell, Nurse Tutor

Sir Henry Osmond-Clarke, KCVO, CBE, FRCS
Honorary Consultant Orthopedic Surgeon, The London Hospital

Mr George Pinker, FRCS, FRCOG
Consultant Obstetrician, St Mary's Hospital, London

Dr Neil Pride, MD, FRCP
Senior Lecturer, Royal Postgraduate Medical School, London

The Simon and Schuster Handbook of Anatomy and Physiology
was edited and designed by
Mitchell Beazley Publishers Limited
Mill House, 87–89 Shaftesbury Avenue
London W1V 7AD

Editor	Lawrence Clarke
Art Editor	John Ridgeway
Assistant Editor	Cynthia Hole
Editorial Assistant	Rosemary Mendus
Production	Julian Deeming
Indexer	Valerie Nicholson

© 1978 by Mitchell Beazley Publishers Limited
Text © 1978 James Bevan
Artwork © 1974, 1976, 1977, 1978
by Mitchell Beazley Publishers Limited:
© Rand McNally USA 1976
© 1972 by International Visual Resource
All rights reserved
including the right of reproduction
in whole or in part in any form
Published by Simon and Schuster
A Division of Gulf & Western Corporation
Simon & Schuster Building
Rockefeller Center
1230 Avenue of the Americas
New York, New York 10020

Printed and bound by Mandarin Offset, Hong Kong
1 2 3 4 5 6 7 8 9 10

Library of Congress Cataloging in Publication Data
Bevan, James Stuart
The Simon and Schuster Handbook of Anatomy and
Physiology
Includes index.
1. Human physiology – Pictorial works.
2. Anatomy, Human – Pictorial works. I. Title.
II. Title: Handbook of Anatomy and Physiology
QP38.B48 1979 612 78-20842

ISBN 0-671-24959-2

Foreword

Not long ago, I was fluoroscoping a patient's chest. (For those of you who don't remember what this involves, you stand behind a movable screen in a darkened room, while the doctor shifts the screen from side to side across your chest, watching your heart beat, seeing your diaphragm move up and down with each breath.) As I peered through the screen, my patient said to me, "I've been having some trouble with my lower bowel. Can you see any of it up there?" This is no exaggeration; it actually happened. Nor was the man illiterate or uneducated. He had read many books on medicine for the layman. They all told him what certain symptoms meant and how he might prevent or treat them. He read all about the dangers of high blood pressure, that there are one million heart attacks each year in the United States alone, that too much alcohol can give liver disease, that an enlarged prostate keeps you running all night. But he didn't know where the bowel is— and probably didn't know about arteries, their location, and what high blood pressure does to them. Nor could he visualize what happens in a heart attack, or what actually takes place when the liver develops cirrhosis.

No matter how many medical books and magazines and articles you read, you can't really understand health and disease, prevention and treatment, unless you are able to visualize the particular organ within the body, know how it works, and how blood vessels nourish it and nerves activate it. You must appreciate structure in order to understand function. And that is what this marvelous illustrated handbook is all about: an enlightening combination of structure and function. As a result, it is invaluable to anyone interested in medicine at any level—layman, biology student, nurse, paramedic, medical technician, pre-med student, and even the doctor who wants to refresh his memory about anatomy and physiology in areas outside his own specialty.

Dr. Bevan is a gifted physician and surgeon. With a team of superb medical artists, he has achieved something that is very difficult to do. He has diagrammed all the systems of the human body so that any interested reader can understand them. This has been done with accuracy, yet without the unnecessary and complicated minutiae which only turn off the lay reader. Superb perspective and color, careful attention to scale, and an emphasis on how things work in every living organ and system make this book an invaluable companion to any home medical guide.

Open this handbook to any page. Everything you need to know about the illustration is on that same page. What a relief not to have to keep flipping back and forth as you try to correlate text and diagram. And then, how practical it is to have unfamiliar terms defined right there in a column on the side of the page, not buried in a glossary somewhere so that searching for it distracts you and interrupts your train of thought. Wondering about lipoproteins as you try to unravel the mysteries of digestion? Just look to the left, and there it is: "Combination of fat and protein formed in the liver as a method of transport to the parts of the body." Not a wasted word and exactly what you need to know. If you've been reading elsewhere about high density lipo-proteins and are worried that yours are too low, you can at least appreciate, from the definition, what it is and, from the text and illustration, how the protein attached to the fat circulates in the bloodstream as a potential source of energy.

And so it is for virtually any disease you want to understand or any operation you may need to have in the near future. Are you planning to have a vasec-tomy? How does the vasectomy relate to your sperm production, and why will cutting the vas deferens leave the rest of your reproductive and hormonal system intact? Just look at page 76.

Do you wonder why a coronary bypass operation is not really "open-heart" surgery? Turn to page 30 and see the coronary arteries as they course on the surface of the heart.

Dr. Bevan and his team of medical experts have rendered a service to the lay public and to the profession itself in preparing this superior work. If a picture is, indeed, worth a thousand words, then this combination of superb illustrations and illuminating text will take its deserved place on many home and library bookshelves.

Isadore Rosenfeld, M.D.
New York Hospital—Cornell Medical Center

Medical Terms 1

A great many apparently difficult medical terms are constructed from comparatively few elements. Once these elements are recognized many pieces of medical jargon read quite naturally and easily.

Element	Meaning	Example
a	without	a-morphous: without shape
ab	away from	ab-duct: to lead away from
ad	towards	ad-renal: towards the kidney
aden, adeno	gland	aden-itis: inflammation of a gland
aesthes	see esthes	
alg, alge	pain	an-alge-sia: without pain
an	without	an-algesia
andro	male	andro-gen: of male origin
angi, angio	vessel	angio-gram: a record of a blood vessel, usually made by special X-ray technique
ante	before	ante-natal: (examination carried out) before the birth
arter, arterio	artery	arterio-sclerosis: hardening of the arteries
arthr, arthro	joint	arthr-itis: inflammation of joints
bil, bili	bile	bili-ary: concerning the biliary system
brachi	arm	brachi-al: concerning the arm
brady	slow	brady-cardia: a condition of slowing of the heart
bronch, bronchi	windpipe	bronchi-tis: inflammation of the windpipe
carcin, carcino	cancerous	carcino-genic: something that causes cancer
cardi, cardio	heart	cardio-myopathy: a sickness of the muscle of the heart
cele	swelling	cysto-cele: swelling of a bladder
cephal, cephalo	head	a-cephal-ous: without a head
cervic	neck	cervic-itis: inflammation of the neck of a structure
chol, chole	bile	chol-angitis: inflammation of the bile vessels
chondr, chondri	cartilage	chondr-oma: swelling of the cartilage
cortex, cortico	outer part	cortico-trophin: something that causes growth of the cortex
cost	rib	cost-al margin: the edge of the ribs
crani, cranio	skull	cranio-tomy: cutting the skull
cryo	freezing	cryo-surgery: surgery using a freezing technique
crypt, crypto	hidden	crypto-genic disease: disease in which the origin is hidden
cut	skin	cut-icle: thin skin
cyst	bladder	chole-cyst-ectomy: the removal of the gall bladder
cyt, cyto	cell	nucleo-cyt-e: cell with a nucleus
dent	tooth	dent-ist
derm, dermato	skin	intra-derm-al: within the skin
dis	undo	dis-organize: prevention of the normal working
dors, dorso	back	dorso-lumbar: back and lumbar region
dys	bad, disorder	dys-entery: disorder of the intestine
ectomy	cut out	gastr-ectomy: removal of the stomach
edem	swelling	lymph-edem-a: swelling due to lymph
en	in	en-cephalitis: inflammation of the inside of the brain
end, endo	inside	endo-scopy: to look inside
enter	intestine	gastro-enter-itis: inflammation of the stomach and intestine
erythro	red	erythro-cyte: red cell
esthes	feel	an-esthes-ia: without feeling
ex	out of	ex-hale: to breathe out
extra	beyond	extra-systole: a systole beyond the normal ones
form	shape	lenti-form: shaped like a lentil
gastr, gastro	stomach	gastr-ectomy: removal of the stomach
gen	originate	cyto-gen-ic: originating within the cell
gloss, glosso	tongue	glosso-pharyngeal: concerning the tongue and pharynx
gram	recording	cardio-gram: a record of the heart
graph	written record	encephalo-graph: a written record of the inside of the brain
gyn, gyneco	woman	gyneco-mastia: excessive development of male breasts
hem, hemat, hemo	blood	hemat-uria: blood in the urine
hemi	half	hemi-plegia: half of the body that is struck by a condition, stroke
hom, homeo, homo	same	homo-genous: the same origin
hydr, hydro	water	hydro-phobia: abnormal fear of water
hyper	above	hyper-thyroidism: overactive thyroid gland
hypo	below	hypo-glossal: below the tongue
hyster	womb	hyster-ectomy: the removal of the womb
iatro	physician	iatro-genic: originating from the physician
in	1. in	in-fusion: to run a liquid into the vein
	2. negative	in-firm: no longer strong
infra	below	infra-red: below the wavelength of red light
inter	between	inter-costal: between the ribs
intra	inside	intra-venous: inside the vein
is, iso	equal	iso-metric: equal length
itis	inflammation	nephr-itis: inflammation of the kidney
kin, kine	movement	hyper-kinetic: overactive movement
lab, labi	lip	labi-a majora: large lips
lact, lacto	milk	lacto-genic: causing the formation of milk
larynge, laryngo	windpipe	laryngo-scope: to look at the larynx
lent, lenti	lentil-shaped	lenti-form: formed like a lentil
lep, leps, lept	seizure	narco-leps-y: sudden onset of sleep
leuco, leuko	white	leuko-cyte: white cell
lingua	tongue	lingua-l nerve: the nerve to the tongue
lip	fat	lip-oma: a fatty swelling
lymph	water	lymph-edema: swelling due to water
macro	large	macro-cyte: large cell
mal	bad	mal-ady: illness
man, manu	hand	manu-al: using the hand
mani	mental disorder	hypo-mani-c: below the stage of mental illness
mast	breast	mast-ectomy: removal of the breast

Medical Terms 2

Term	Meaning	Example
mega, megalo	great	megalo-mania : mental disorder involving delusions of greatness
men, mens, meno	month	dys-meno-rrhoea : disorder of the menstrual flow
metr, metro	womb	endo-metr-itis : inflammation of the lining of the womb
micro	small	micro-cephaly : a small head
mon, mono	single	mon-ocular : one eye
mot	movement	oculo-mot-or nerve : a nerve that controls the eye muscle
my, mya, myo	muscle	my-algia : pain in the muscle
myel, myelo	marrow, centre	myelo-cyte : a cell from the marrow
nas, naso	nose	naso-pharynx : nose and pharynx
necro	death	necro-tic : concerned with death
neo	new	neo-plasm : new shape
nephr	kidney	nephr-itis : inflammation of the kidney
neur, neuro	nerve	neuro-logy : the knowledge of the nervous system
norm, normo	usual	normo-chromic : normal colour
nutri	nourish	mal-nutri-tion : badly nourished
ocul, oculo	eye	ocul-ist : eye specialist
odont	tooth	orth-odont-ist : one who straightens teeth
odyn	pain	pleur-odyn-ia : pain of the side
oid	form	andr-oid : formed like a man
olig, oligo	small, few	oligo-menorrhoea : a small amount of menstrual fluid
oma	swelling	oste-oma : swelling of the bone
oo	of eggs	oo-phoritis : inflammation of the ovary
ophthalm, ophthalmo	eye	ophthalmo-scope : an instrument for looking at the eye
or	mouth	or-al : concerning the mouth
orchi	testis	orchi-dectomy : removal of the testis
orth, ortho	straight	orth-optics : straightening of vision
oss, osteo	bone	oss-icle : small bone
ot, oti	ear	ot-itis : inflammation of the ear
ov	egg	ov-ary : female sex organ where eggs are formed
par	bear	multi-par-ous : one who has carried many children
par, para	at the side of	para-medical : someone who works beside doctors
paed, ped	child	ped-iatrics : the treatment of disease in children
		ortho-ped-ics : the correction of deformities in children
path, patho	sickness	patho-logy : the knowledge of sickness, in this case from studying dead tissue
pen, penia	lack	thrombocyto-penia : lack of clotting cells
peri	around	peri-cardium : that which is around the heart
phag, phago	eat	eso-phag-us : the gullet
pharyng, pharynge	throat	glosso-pharynge-al nerve : the nerve to the tongue and throat
phleb	vein	phleb-itis : inflammation of a vein
phob	fear	hydro-phob-ia : fear of water
phon, phono	sound	phono-cardiograph : writing from the sounds of the heart
phor	carrier	oo-phor-itis : inflammation of the ovary
plas, plasm, plast	shape	cyto-plasm : the shape of the contents of the cell
plegi	stroke	hemi-plegi-a : a half-stroke
pleur, pleuro	rib, side	pleur-odynia : pain in the side
pneum, pneumon	lung	pneumon-ectomy : removal of the lung
poie	produce	hemo-poie-tic : blood producing
poly	many	poly-arthritis : inflammation of many joints
post	behind	post-pharyngeal : behind the pharynx
pre	before	pre-molar tooth : tooth in front of the molar tooth
pro	before	pro-geria : early ageing
proct, procto	anus	procto-scope : an instrument for looking inside the anus
pseud, pseudo	false	pseudo-cyesis : false pregnancy
psych, psycho	mind	psycho-logy : knowledge of the mind
py, pyo	pus	pyo-genic : concerning the formation of pus
radi, radio	radiation	radio-logy : study of radiation
ren	kidney	supra-ren-al gland : the gland which rests above the kidney
retro	behind, backwards	retro-grade amnesia : loss of memory of what occurred before an accident
rhage	flow	hemo-r-rhage : flowing of blood
rhea	gush	pyo-r-rhea : gushing of pus
rhin, rhino	nose	rhin-itis : inflammation of the nose
salping	tube	salping-itis : inflammation of the uterine tube
sclero	hard	arterio-sclero-sis : hardening of the arteries
scop	look at	ophthalmo-scop-e : an instrument to look at the eyes
sect	cut	dis-sect : to cut apart
semi	half	semi-circular canal
sep, seps, sept	decay	seps-is : destructive infection of tissue by bacteria
splanchn	intestine	splanchn-ic artery : the artery that goes to the intestine
steat, steato	fat	steato-rrhea : fatty diarrhea
stom	mouth	colo-stom-y : an opening, mouth, in the colon
sub	below	sub-mandibular : below the mandible
super, supra	above	supra-umbilicus : above the umbilicus
sy, syl, sym, syn	with	syn-chronous : at the same time sym-biosis : living together
tachy	fast	tachy-cardia : fast heart-beat
tele	distant	tele-cardiography : taking a cardiograph at a distance
therap	treatment	radio-therap-y : treatment with radio-active substances or X-rays
thorac, thoraco	chest	thoraco-tomy : cutting into the chest
thrombo	clot	thrombo-phlebitis : inflammation and thrombosis of the veins
tom	cut	lobo-tom-y : cutting of the frontal lobes
tracheo	windpipe	trache-o-stomy : an opening, mouth, in the windpipe
troph	growth, nurture	dys-troph-y : disorder of growth
ultra	beyond	ultra-sonic : beyond the normal range of sound
uri	urine	poly-uri-a : a condition of frequent passing of urine
vesic	bladder	retro-vesic-ular : behind the bladder

Glossary 1

There are certain conventions when describing the position of a part of the body. It is assumed that the individual is standing upright with the palms of the hands facing forwards. This means that the diaphragm is superior to the liver and is inferior to the lungs, and the thumb is on the lateral side of the hand. There is a more detailed glossary of terms under the heading "Definitions" in each section of the book, from "How the Cell Works" to "Metabolism and Nutrition". Diseases are defined under the heading "Common Diseases".

Abdomen   Lower part of the trunk between the diaphragm and the pelvic girdle.

Absorption   Process by which substances cross through the mucous membranes or skin. The products of digestion are absorbed from the small intestine into the blood-stream.

Acoustic   Anything that is associated with sound or the sense of hearing.

Adipose   Of a fatty nature.

Adrenaline   Hormone produced by the medulla of the adrenal gland.

Agglutination   Clumping together of cells in fluid.

Alimentary canal   Digestive tract, from the mouth to the anus.

Allergy   Exaggerated response by the body's immune system to a specific external antibody, for example pollen, which does not produce a response in the majority of people.

Amino-acid   One of the basic units from which proteins are made.

Anabolism   Building of complex substances from simple components. The opposite to catabolism.

Anastomosis   Joining together of two pieces of tissue, for example, the joining of the jejunum to the stomach after a partial gastrectomy.

Anatomy   Study and knowledge of the basic structure of man or animals.

Androgen   Any substance that produces results similar to that of the male hormone.

Anesthetics   Use of drugs to make a patient unaware.

Anterior   Front or forward part of the body.

Antibiotic   Substance either made from a fungus or living micro-organism or synthesized; it has the ability to kill or prevent the growth of pathogenic organisms.

Antibody   Specific substance produced by the body as a reaction to and in an attempt to dispose of a foreign substance—antigen.

Antigen   Any substance which, when introduced into the body, causes the formation of an antibody.

Antiseptic   Any chemical used to destroy harmful micro-organisms.

Antitoxin   Antibody that neutralizes bacterial poisons—toxins.

Arteriosclerosis   Loss of elasticity of arterial walls with deposition of cholesterol in the degenerating areas.

Arthrology   Study of joints.

Articulation   The action or movement of a joint.

Aseptic   Area that is free from infective organisms.

Aspirate   To remove fluid or air from a cavity within the body.

Auditory   Concerned with the mechanism of hearing.

Bacteria   Microscopic organisms which are classified by their shape and staining capabilities. They live in many areas of the natural environment and only a few of them are pathogenic to the human body.

Basal metabolism   Minimal rate at which the resting body uses energy.

Benign   Not malignant.

Bifurcation   Forking or division into two branches.

Biochemistry   Study of the chemistry of living processes.

Calibration   Measurement of an object.

Calorie   Amount of heat required to raise 1 gramme of water through 1°C. The calorie value of food is the number of calories it would yield, as energy, if completely used. Nutritionally the Kilocalorie is used and is written as 1 Calorie, with a capital C.

Cancer   Malignant growth in which the cells develop in a way that is no longer controlled and may spread throughout the body.

Carbohydrate   Basic structure of starch or sugar.

Carcinogen   Substance which will cause cancer to develop.

Cardiology   Study of the heart.

Catabolism   Breakdown of complex body substances into simpler compounds.

Cellulose   Complex carbohydrate that forms a basic structural material of all plants.

Cephalic   Concerned with the head.

Cervical   Concerned with the neck or cervix.

Cervix   Literally means neck, may mean cervix of uterus, tooth or bladder.

Chromosome   Thread of genetic material. One of 23 pairs in the cell nucleus.

Chyme   Liquid food in the stomach or small intestine.

Cilia   Small hair-like structures on the outer surface of cells.

Coitus   Sexual intercourse.

Conception   Moment of fertilization.

Consciousness   State of being awake.

Corpuscle   See *Red blood corpuscle* and *White blood corpuscle*.

Cortex   Outer layer, for example of the brain, adrenal gland or kidney.

Cranial   Anything concerned with the skull.

Cytology   Microscopic study of cells.

Cytoplasm   Clear jelly-like substance that supports the intracellular contents.

Decibel   Unit of measurement of sound.

Dermatology   Study of skin disease.

Diet   Food and drink that is consumed.

Diffusion   Movement of molecules in a fluid from an area of high concentration to one of low concentration.

Dilation   Stretching or widening.

Dissemination   Scattering of a substance over a wide area, for example, cancer is disseminated throughout the body.

Distal   Some distance from a certain place.

Glossary 2

Diuretic   Any substance that increases the output of urine.

Dorsal   Anything that is related to the back.

Embryo   Fetus in the early stages of development, before the fourth month.

Endocrine gland   Gland that secretes a hormone directly into the blood stream.

Endocrinology   Study of the endocrine glands.

Enucleate   To remove an object from its place in the surrounding tissue without damage.

Enzyme   Substance which speeds up chemical reactions.

Estrogen   One of the group of female sex hormones which is principally produced by the ovary before ovulation but also from the adrenal glands.

Etiology   Origin and cause of a disease.

Exacerbation   Periods when a disease, or pain, becomes worse.

Exocrine gland   Gland that secretes substances into a duct leading to the area where it is used; for example, the salivary gland is an exocrine gland.

Fascia   Any sheet of thin fibrous tissue.

Feces   Semi-liquid contents of the large intestine.

Fertilization   Moment when the sperm enters the ovum.

Fetus   Unborn infant from about the twelfth week of pregnancy until the time of birth.

Fissure   Term for any crack or cleft in the body.

Frontal   That which is in front, particularly concerned with the forehead.

Fructose   Simple sugar found in fruits.

Gastric   Anything that is concerned with the stomach.

Gastro-enterology   Study of the alimentary canal.

Genital   Anything that is concerned with the sexual organs.

Genito-urinary disease   Study of disease of the urinary and reproductive systems.

Geriatrics   Care of the elderly.

Gestation   Period of time between conception and birth.

Gynecology   Study of women's diseases.

Hematology   Study of the blood cells.

Hemopoiesis   Production of blood cells.

Hepatic   Anything concerned with the liver.

Heredity   Passage of family characteristics from parents to child.

Homogenous   Same consistency all the way through.

Hormone   Chemical produced in one part of the body, an endocrine gland, and carried to another area where it has a specific effect.

Illusion   Incorrect reception of something actually present.

Immunity   Body's ability to resist infection.

Immunology   Study of the way the body reacts to disease.

Infection   Invasion of the body by pathogenic organisms.

Inferior   Situated below.

Inflammation   Response of tissues to injury or infection. The signs are heat, redness, swelling and pain.

Ingestion   Act of taking substances into an organism, for example, eating food or the ingestion of a bacteria by a white blood corpuscle.

Inhibition   Slowing down or stopping of a reaction, either chemical or neurological.

Intelligence   Individual's capacity to think rationally, learn and react to the environment.

Ischemic   Insufficient blood supply to a part of the body.

Isometric   Means the "same length" and is used medically to mean contraction of a muscle without bending a joint.

Lateral   Concerning the side.

Libido   Term used by Freud to describe the instinctive sexual drive in men and women.

Malignant   Tendency to go from bad to worse, particularly used in reference to cancer.

Mastication   Chewing of food.

Maturation   State of full development.

Medial   Towards the mid-line.

Medulla   Term for the central part of an organ, for example, the medulla of the adrenal gland.

Meiosis   Cell division in which there is a splitting to form a gamete with halving of each chromosome to form a chromatid.

Menarche   Onset of menstruation at puberty.

Menopause   Final cessation of menstruation in women usually occurring around the middle or late forties.

Menstrual cycle   Sequel of events that occurs due to hormonal changes beginning and ending with the onset of menstruation. This usually takes about 28 days.

Menstruation   Monthly "period" of uterine bleeding that occurs when the lining is shed.

Metabolic rate   Rate at which an individual consumes energy reflected directly in the body's oxygen intake.

Meticulous   With great care and precision.

Micturition   Act of urination— passing water.

Mitosis   Division of a cell to produce an identical one which contains the same number of chromosomes.

Motile   Something that moves by itself.

Mucous membrane   Thin, delicate lining, found in many of the body's internal surfaces, which is moistened with mucus secreted by the mucous glands in the membrane.

Mucus   Semi-liquid substance secreted by the mucous glands in the mucous membranes.

Myology   Study of muscles.

Nasal   Concerning the nose.

Necrosis   Death of a tissue.

Neurology   Study of the nervous system.

Neuropathy   Degeneration of a nerve.

Obstetrics   Care and delivery of the pregnant woman.

Occipital   Anything to do with the occiput—the back of the skull.

Olfactory   Anything concerning the sense of smell.

Oncology   Study of cancer.

Ophthalmology   Study of the eye and sight.

Optic   Anything to do with the eye and sight.

Organism   Living thing, whether animal, plant or microbe.

Orthopedics   Study of bone disease.

Osmosis Selective passage of the molecules of a liquid through a semipermeable membrane between solutions of different concentrations. Water from a weaker solution moves through the membrane to dilute a stronger one.

Osmotic pressure Pressure required to stop water from passing through a membrane into a solution; for example, osmotic pressure occurs when fluid is forced back into the capillary system after it has been forced out into the body tissues by the blood pressure.

Osteology Study of bones.

Otorhinolaryngology Study of the ear, nose and throat.

Ovulation Release of a mature ovum from an ovary.

Palmar Anything concerning the palm of the hand.

Palpation To examine a patient with the hand.

Parturition Act of giving birth to a child.

Patent Open.

Pathogenic That which gives origin to a disease, for example, a virus is the pathogenic organism of influenza.

Pediatrics Study of children's diseases.

Peptone Partly digested protein.

Perception Overall awareness of the environment from the interpretation of sensory information.

Peripheral Towards the outside.

Pharmacology Study of drugs in relation to medicine.

Physiology Understanding of the workings and functions of living organisms.

Plantar Concerning the sole of the foot.

Posterior Towards the back or situated behind.

Pregnancy Time from the moment of fertilization to parturition.

Progesterone Hormone produced by the ovary in the second part of the menstrual cycle to maintain the endometrium during implantation of the blastocyst.

Pronation Placing or holding of the palm of the hand downwards.

Prophylactic Something that is given to try and prevent a disease occurring.

Protein Complex nitrogen-containing substance concerned with building the body's fabric.

Protuberance Bulge or swelling.

Proximal Close to a certain place.

Psychiatry Study of mental illness.

Puberty Period during which the sexual organs reach maturity with the development of the secondary sexual characteristics, such as growth of hair, breast development and change of voice.

Quickening When the fetus is first felt to move by the mother is the onset of quickening.

Radiology Study of X-rays.

Radiotherapy Treatment of disease using X-rays.

Red blood corpuscle Small hemoglobin carrying cell that does not have a nucleus.

Renal Anything concerned with the kidney.

Reproductive system Sex organs concerned with producing sperm, the ovum and maintaining a pregnancy once it has started.

Respiration Process by which oxygen is delivered to the tissues and carbon dioxide removed.

Retrosternal Behind the sternum.

Rheumatology Study of the diseases of muscles and joints.

Sac Cavity within the body tissues, for example the pericardial sac.

Sclerosing Hardening of a tissue.

Sensation Experience of stimulation of a sensory nerve.

Serum Clear yellow fluid that separates from clotted blood.

Skeleton Articulated bony framework of the body.

Smell Response of the olfactory nerve endings to chemical stimulation.

Stenosis Narrowing, for example of an artery.

Superior Above or on top of.

Supination Placing or holding of the palm of the hand upwards.

Synovial membrane Smooth membrane that secretes a small amount of fluid and usually forms a sac to protect areas where friction may occur, for example joints and around tendons that pass over joints.

Taste Sensation of flavour that occurs when the sensory areas of the tongue are stimulated.

Tendon Strong, fibrous bundle which joins a muscle to a bone.

Thorax Chest compartment enclosed by the ribs, backbone and diaphragm. It contains the lungs, the heart, major blood vessels and the esophagus.

Threshold Critical level of a stimulus to produce a response; for example, a feather lying on the skin may not be noticed, but a piece of gravel is above the threshold for feeling.

Thrombosis Formation or development of a clot of blood.

Thyroid gland Endocrine gland in the neck which produces hormones to stimulate the body's metabolism.

Tissue Large areas of similar cells, for example muscle tissue.

Torsion Act of twisting.

Traction Action of pulling on a muscle or organ by means of weights to correct an abnormal condition.

Transient Something that occurs only for a short time.

Trauma Wound or injury; emotional upset that may lead to mental illness.

Tremor Trembling.

Umbilical Concerning the navel.

Unconscious Absence of normal wakefulness, the lack of awareness that occurs when there is no response to normal stimuli. It is not, as a rule, applied to natural sleep.

Venereology Study of sexually transmitted diseases.

Ventral Concerning the abdomen.

Virology Study of virus disease.

Virus Complex chemical dependent on living cells for its reproduction.

Viscus Usually means the bowel, but can be referred to any of the abdominal organs, for example the liver and spleen.

Vitamin One of a series of substances that are found in the normal diet and are required in small amounts for the normal metabolism of the body and which the body cannot make from raw materials because it lacks the enzymes needed for their synthesis.

White blood corpuscle Concerned with fighting infection in the body.

The Whole Body 1

Looking at the body

The systems that make the body work are all interdependent. This is sometimes difficult to remember when it is a matter of convenience to study each system separately. It makes learning about them easier, but it is just as possible to approach the body in a different, unconventional manner.

Little is known of how people develop as individuals with their own physical and psychological differences. Some of these are the result of inheritance; clearly much is due to immediate family, and from reactions to the wider world. Brain functions are difficult to measure; even intelligence tests are often inaccurate and may give little indication of the brain's power of original thinking.

However, the study of a particular protein will tell a great deal about how the body functions normally. Its reaction with the digestive juices breaks it into the constituent parts, amino-acids, which are then absorbed through the intestine wall and into the blood stream to the liver. In the liver the amino-acids may be rebuilt into new proteins or sent by the arterial system to help replace old proteins, such as those in muscle or bone. These "worn out" amino-acids are removed from the tissues by the venous system and returned to the liver. Here they can be rebuilt or broken down into urea; urea is excreted by the kidneys into the urine.

This unconventional way of learning teaches how the digestive tract, circulation, muscle and metabolism work, but could easily involve the hormones from the endocrine system, which help instruct the body, or the nervous system, where protein may be required.

A great deal is now known about how the brain controls the other systems of the body, either through quick direct nervous response or secretion of chemicals—hormones—to produce a slower reaction. The way in which the balance of salts in the blood is maintained by foods that are eaten is understood. All these physical occurrences can be watched and measured with accuracy.

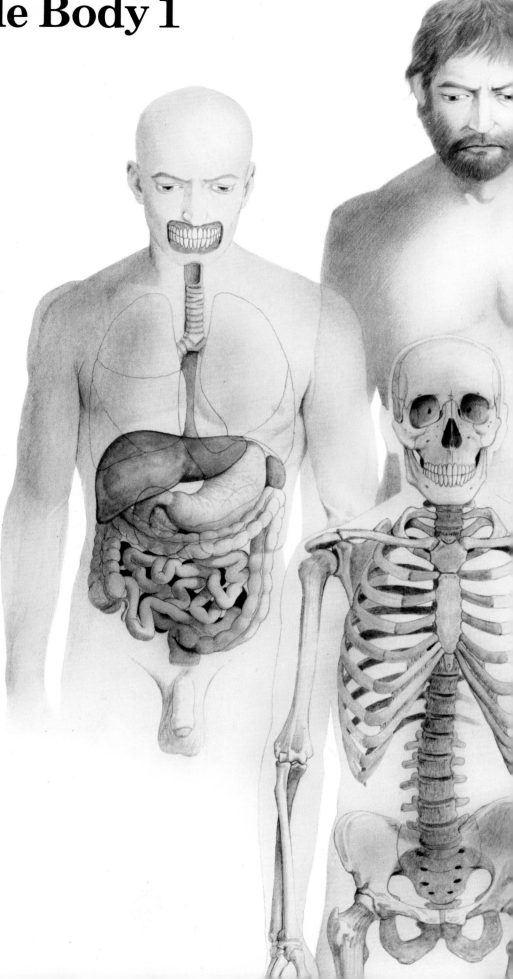

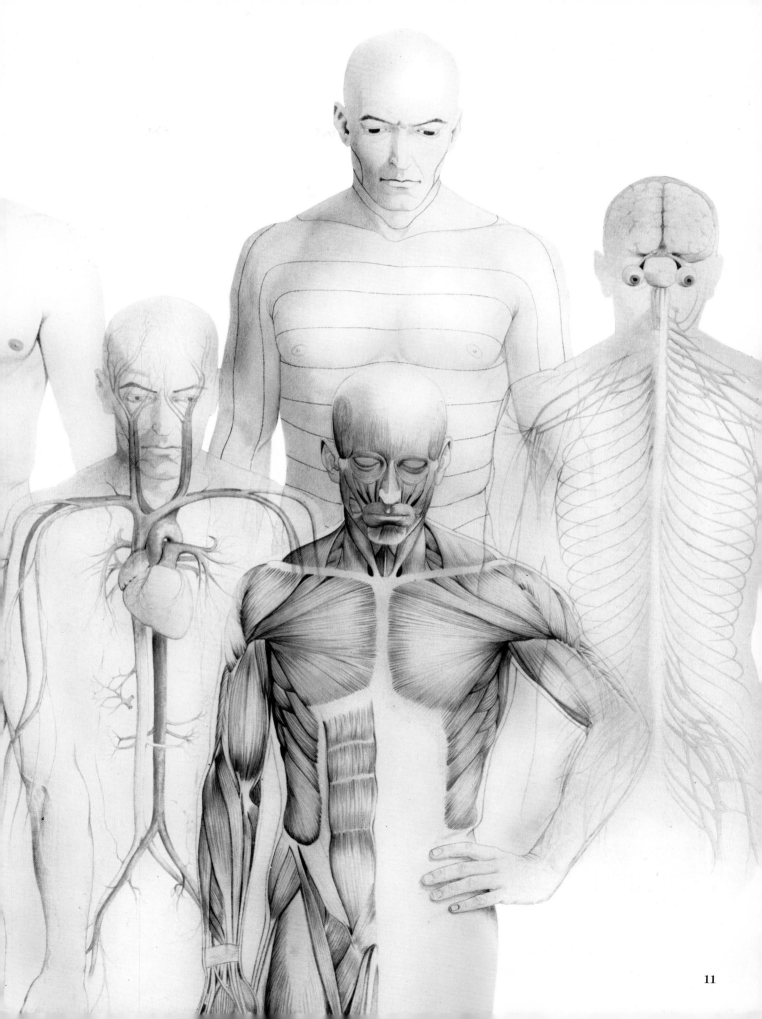

The Whole Body 2

Definitions

Achilles' tendon Large tendon from the calf muscles to the heel bone—the calcaneus.

Amino-acid One of the basic units from which proteins are made.

Antibody This is a specific substance produced by the body as a reaction to and an attempt to dispose of a foreign substance—antigen.

Aponeurosis Tendinous expansion connecting a muscle with the parts that it moves.

Areolar tissue Loosely interwoven connective tissue supporting organs and structures.

Artery Thick-walled blood vessel carrying blood under pressure from the heart.

Cartilage Special tough form of connective tissue, the degree of hardness depending on the number of collagen fibres.

Collagen fibre Strongest form of connective tissue.

Connective tissue Supporting tissues of the body.

Elastic fibre Elastic form of connective tissue.

Fat cell Cell specialized for the storage of fat to be used as reserves for energy or insulation.

Ground substance Jelly-like material of salts and water, protein and carbohydrate that fills in the gaps of many of the areas of connective tissue.

Inheritance Passage of family characteristics from parents to child.

Ligament Tough, fibrous band connecting bones or supporting organs.

Macrophage Large tissue cell, either free or attached, that engulfs bacteria.

A general description

The whole body is an amazing compromise between rigidity and mobility. The internal organs (*right*) are closely packed together and yet can work freely and easily. The surrounding framework of bone and muscles applies support and protection.

The skeleton gives the upright strength to the body and in some places, such as the skull and thorax, acts as a protective layer. The joints give the bones mobility and the muscles strength and suppleness.

The contents of the chest and abdomen are constantly moving—the beating of the heart, inspiration and expiration in the lungs and peristalsis of the bowel. These structures can move without difficulty as they are surrounded by special, smooth layers of tissue known as pericardium, pleura and peritoneum. These small cavities of tissue, or sacs to give them their medical name, are rather like a very soft balloon containing a little fluid. Push a clenched fist in one side of a balloon, press the balloon against the palm of the other hand; the fist moves easily and without friction. Similar "sacs" occur in the joints, around tendons passing over joints and at the points of friction, such as the knee and elbow. These are known as synovial sacs.

The largest organ in the body is the skin. In an adult it covers about 2 square metres and not only envelops the whole body in a protective waterproof layer but is also part of the heat-regulating system. The liver is the most complicated organ with the greatest number of functions, transforming digested food into usable materials and disposing of waste substances.

The circulation is constantly restoring and revitalizing as well as removing waste products from the basic unit of the body—the cell. The cell is a microscopic structure of which there are billions that build up the whole body. Each cell specializes and carries out its own particular function. All the structures and organs are held together by the connective tissue, made up of cells that act as a kind of packing to protect and support the internal mechanisms.

The internal organs

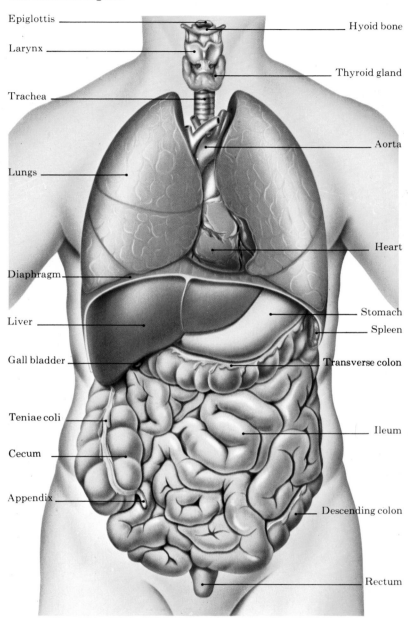

Epiglottis — Larynx — Trachea — Lungs — Diaphragm — Liver — Gall bladder — Teniae coli — Cecum — Appendix — Hyoid bone — Thyroid gland — Aorta — Heart — Stomach — Spleen — Transverse colon — Ileum — Descending colon — Rectum

The contents of the body

The skeleton: The axial and appendicular skeletons.

Joints; muscles; ligaments.

Heart; arteries; veins; blood; lymph vessels; bone marrow; spleen.

Respiration: Nose; sinuses; trachea; 2 bronchi; 2 lungs.

Digestion: Mouth; esophagus; stomach; small intestine; colon; liver; gall bladder; pancreas.

Urinary system: 2 kidneys; 2 ureters; bladder; urethra.

Endocrine glands: Pituitary; thyroid; 4 parathyroids; 2 adrenal; pancreas; 2 testes or ovaries.

Nervous system: 2 eyes; 2 ears; organs of hearing and balance, smell and taste; brain; spinal cord; peripheral nerves; autonomic nervous system.

Skin, hair and nails.

Reproduction: Female—2 ovaries; 2 Fallopian tubes; uterus; vagina. Male—2 testes; 2 epididymi; 2 sperm ducts; prostate.

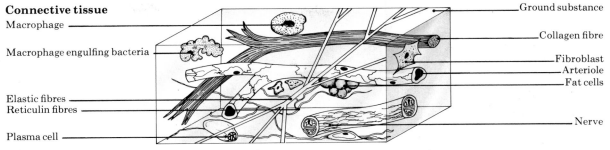

Connective tissue
Macrophage

Macrophage engulfing bacteria

Elastic fibres
Reticulin fibres

Plasma cell

Ground substance

Collagen fibre

Fibroblast
Arteriole
Fat cells

Nerve

Connective tissue
The skeleton keeps the organs, blood vessels and nerves in place and, to a certain extent, is a protection. The connective tissue supports and binds them together (*above*). It also supplies the ligaments and tendons for the joints and muscles, the tethering for the larger organs, the softness for protection and rigidity in the form of cartilage.

There are many forms of connective tissue, but they are all developed from the same jelly-like "ground substance", made up of salts and water, protein and carbohydrate. Embedded in this jelly are the various fibres and cells: Elastic fibres to give elasticity; collagen fibres to give strength; reticulin fibres to give support; white cells and macrophages to fight infection; fat cells for storage; plasma cells to produce antibodies.

Collagen fibres—tendon

Tendons and ligaments
The strongest connective tissue is found in the tendons and ligaments (*above*). These are made of densely packed collagen fibres and may be very thick, as in the Achilles' tendon, or thin and wide, as in the thin sheet of the aponeurosis that covers the skull and into which many of the skull muscles are inserted. They have to resist the pull of muscles and mobility of joints with suppleness and strength, without elasticity.

Areolar
Areolar tissue
Arteriole

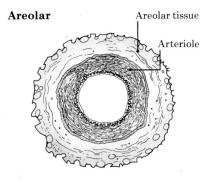

Areolar tissue
The proportion of the different fibres in the "ground substance" gives the connective tissue its variable characteristics.

Areolar tissue is formed throughout the body in loose sheets around blood vessels, nerves and tendons; as a soft, pliable substance it helps to fill the spaces between larger organs. The tissue is made of a mixture of collagen, elastic and reticulin fibres. Under the skin and face it contains a large amount of mobile, elastic fibres, in contrast to the palms of the hands and soles of the feet, which are tough and contain more collagen fibres.

Connective tissue is found holding the cells together within organs. This can be seen in the liver, where the bile ducts, veins and arteries are held within the liver substance and in the artery wall itself; here the circular muscle is held in place by a protective sheath of loose areolar tissue (*above*). Certain diseases known as collagen diseases can afflict connective tissue, causing disorder of their normal function. The skin becomes wrinkled and creased with age as the elastic fibres degenerate. This occurs more rapidly when they are damaged by ultraviolet sunlight which is partly prevented by more pigmentation.

Cartilage
Cartilage is a special form of connective tissue and supplies the fabric for the formation of bone. Bone forms by ossification—minute crystals of calcium salts are manufactured by osteoblast cells and arranged in layers.

Adult cartilage does not contain blood vessels or nerves but is filled with small holes to allow nutrition to seep into it. There are three forms of cartilage: Elastic cartilage is mainly densely packed cells to give it the kind of springiness found in the ear; fibro-cartilage is tough and contains many more collagen fibres (the intervertebral disc of the spine has a thick circle of fibro-cartilage around the softer centre of dense connective tissue, the nucleus pulposus); and hard hyaline cartilage, found at the bone ends and in the nose, is made of dense collagen fibre.

Fat cells

Fat cells
The fat cells (*above*) have three functions: Storage; insulation; protection over certain areas, such as the buttocks, and around various organs, such as the kidneys and heart and in the liver.

Some of the areas of the body consist mainly of fat storage cells. Fat cells develop in infancy and then their total number remains constant for the rest of life. Fat babies become fat adults.

Metabolism Way in which an individual uses energy.

Ossification The process or state of being changed into a bony substance.

Osteoblast Tissue cell that forms bone.

Pericardium Soft sac that surrounds the heart and reduces friction with a small amount of fluid.

Plasma cell Body cell concerned with the formation of antibodies. It is similar to a large lymphocyte.

Pleura Sac-like structure filled with some fluid that lines the inner surface of the thoracic cage and surrounds each lung.

Protein Complex nitrogen-containing substance concerned with building and repairing the body.

Reticulin fibre Supporting fibres of the connective tissue.

Synovial membrane Smooth membrane that secretes a small amount of fluid to protect areas where friction occurs.

Synovial sac Bag of synovial membrane usually occurring at points of friction, e.g. joints and around tendons.

Tendon Strong, fibrous bundle, which joins a muscle to a bone.

Thorax Chest formed by the 12 pairs of ribs and intervening muscles with the sternum in front and spine or column behind.

Urea Nitrogen-containing substance formed from ammonia in the liver.

Vein Thin-walled blood vessel carrying blood under pressure back to the heart.

White blood cells Concerned with fighting and resisting infection, there are two main varieties—polymorphonuclear leukocytes (polymorphs) and lymphocytes.

How the Cell Works 1

Definitions

Adenosine triphosphate (ATP) Chemical stored in the mitochondria that supplies energy for cell metabolism.

Adrenal gland One of the pair of endocrine glands lying above the kidney. It secretes corticosteroids from the cortex and adrenaline from the medulla.

Adrenaline Hormone secreted from the inner part of the medulla of the adrenal gland.

Allele One of a pair of chromosomes or genes.

Amino-acid One of the basic units for protein.

Carbohydrate Basic structure of starch or sugar.

"Carrier" Individual who has a recessive gene for a condition, but does not develop it. Carried from one generation to another, the condition is only reproduced when it is sex-linked or paired with another recessive gene.

Chromatid The "half" chromosome in the gamete when meiosis occurs.

Chromatin Central substance of a cell nucleus that forms chromosomes.

Chromosome Thread of genetic material. One of 23 pairs in the cell nucleus.

Corticosteroid One of several hormones secreted by the adrenal cortex.

"Crossing over" Method by which genes are interchanged between pairs of chromosomes before meiosis.

Cytoplasm Basic intracellular substance that supports the intracellular structures. It contains minute globules of fat and carbohydrates that combine with protein to form glycoprotein.

The basic cell

The cell (*below*) is the basic unit of life. Single-celled organisms exist as many different varieties, for example bacteria and amoebae. Man is highly complex, with many thousands of different cell functions combined in a body of many hundreds of millions of cells. All these cells develop from one cell—the fertilized ovum.

Each cell is surrounded by both a pliable and protective membrane through which substances may pass in both directions. In the centre is the nucleus, which controls the cell's activities. The contents are suspended in the watery cytoplasm, which stores glycoprotein. Films of material—the endoplasmic reticulum—float in the cytoplasm and supply a surface for many of the chemical activities controlled by the Golgi bodies. The energy for all this activity is stored in the many minute mitochondria in the form of adenosine triphosphate (ATP).

The newly prepared substances and waste products are expelled through the membrane into the surrounding interstitial fluid to be taken away in the circulation.

The cell nucleus

The nucleus controls and directs the activities of the cell. It is surrounded by its own membrane—the nuclear envelope—and is filled with densely packed chromatin. Chromatin uncoils, when the cell divides, to form chromosomes. Chromosomes carry the genes that determine the detailed activity of the cell in relation to the needs of the organism. In the nucleus there are nucleoli. These are centres concerned with the formation of nucleic acid necessary for the repair and formation of new cells.

The basic cell

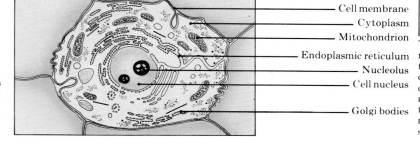

- Cell membrane
- Cytoplasm
- Mitochondrion
- Endoplasmic reticulum
- Nucleolus
- Cell nucleus
- Golgi bodies

The cell as a factory

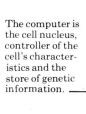

The computer is the cell nucleus, controller of the cell's characteristics and the store of genetic information.

The radar scanner symbolizes the cell's sensitivity to changes in its environment, for the cell can adapt its activities in response to changes both inside and outside itself.

The factory wall, equipped with gates and traffic lights, is the cell membrane, where the entry and exit of molecules are controlled.

The railway takes raw materials into the factory's production plant. The cell takes in small molecules and uses them to construct more complex ones such as proteins.

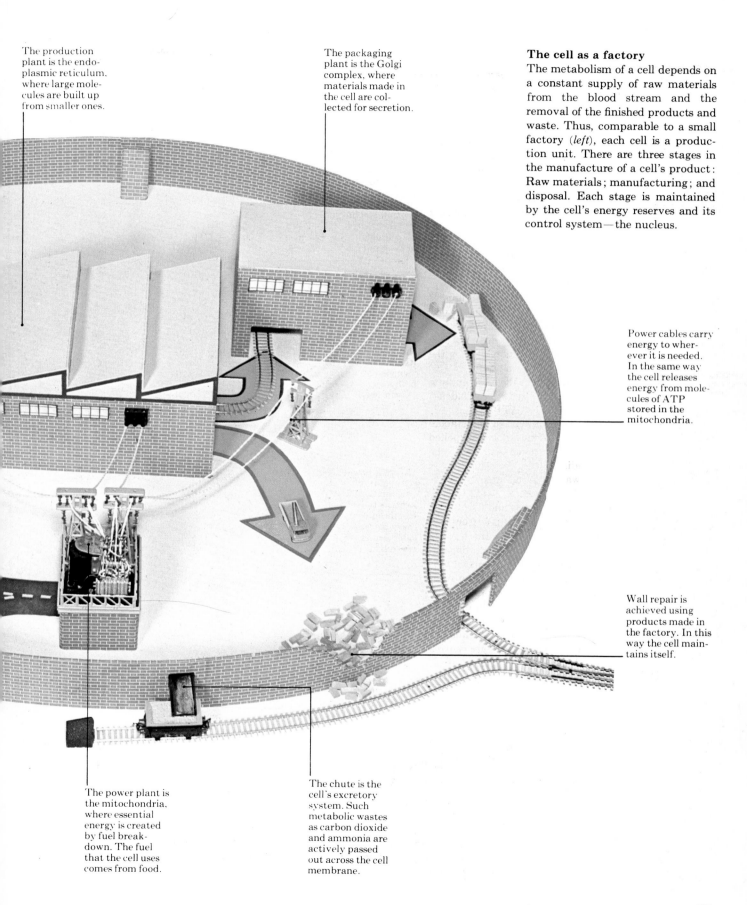

The production plant is the endoplasmic reticulum, where large molecules are built up from smaller ones.

The packaging plant is the Golgi complex, where materials made in the cell are collected for secretion.

The cell as a factory

The metabolism of a cell depends on a constant supply of raw materials from the blood stream and the removal of the finished products and waste. Thus, comparable to a small factory (*left*), each cell is a production unit. There are three stages in the manufacture of a cell's product: Raw materials; manufacturing; and disposal. Each stage is maintained by the cell's energy reserves and its control system—the nucleus.

Power cables carry energy to wherever it is needed. In the same way the cell releases energy from molecules of ATP stored in the mitochondria.

Wall repair is achieved using products made in the factory. In this way the cell maintains itself.

The power plant is the mitochondria, where essential energy is created by fuel breakdown. The fuel that the cell uses comes from food.

The chute is the cell's excretory system. Such metabolic wastes as carbon dioxide and ammonia are actively passed out across the cell membrane.

15

How the Cell Works 2

Cytoplasmic network Faint strands of material that form in the cytoplasm and which pull the chromosomes into the sides of the cell during cellular division.

Deoxyribonucleic acid (DNA) Complex chemical that is the foundation for the memory of the gene. It produces ribonucleic acid.

Dominant The stronger.

Endoplasmic reticulum Film of intracellular matter to which the ribosomes are attached and on which most of the enzyme activities of the cell take place under the direction of ribonucleic acid.

Enzyme Chemical that helps the transformation of one substance into another — a specific catalyst.

Epithelium Layer of cells on the outside of a structure, e.g. skin, or lining the bronchi.

Gamete Cell containing half the number of chromosomes (23) after meiotic cellular division.

Gene Basic unit of a chromosome which carries instructions in the DNA for one body characteristic.

Genotype Combination of genes that produces the same characteristics in an individual.

Glycoprotein Substance formed from the combination of a protein with fat and carbohydrates. It is stored in the cytoplasm.

Golgi body One of a pair of intracellular structures that controls the enzyme activities of the endoplasmic reticulum.

Helix Anything with a spiral form, hence the double helix of the spiralling appearance of DNA.

Heterozygous Chromosomes containing different genes.

The three stages of production in a cell

Raw materials: Each cell type requires different substances and these can only be obtained from the circulation. They are carbohydrates, fats, amino-acids and various salts, which enter the cell through the cell membrane. Depending on the cell's function, only certain substances are allowed through.

Manufacturing: The process takes place on the surface of the endoplasmic reticulum found throughout the cytoplasm. Some of the products may be enzymes or hormones.

Disposal: The final product and the waste substances that are a result of the cell's activities are passed through the cell wall into the interstitial fluid and then the blood for disposal.

The cell's energy reserves

Potential energy is stored in the mitochondria as glucose, fat globules and protein and is released in the form of the chemical adenosine triphosphate (ATP) when it reacts with oxygen. This releases carbon dioxide and water as waste products.

Chemical equilibrium in the cell

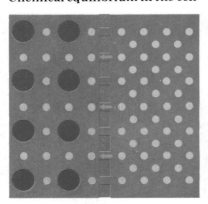

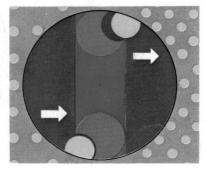

Epithelial cells

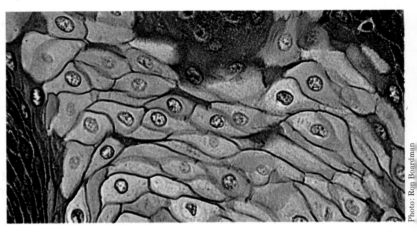

Photo: Ron Boardman

The cell's control

Control for all this activity comes from the nucleus, a memory bank for everything concerned with the cell's functions. The genes in the chromosomes are each a deoxyribonucleic acid (DNA) molecule. They send out chemical messengers in various forms of ribonucleic acid (RNA) to construct the amino-acids into the precise chains to form proteins in the cytoplasm—the material that forms the bulk of the cell.

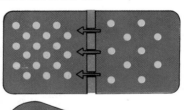

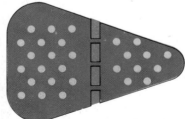

Chemical equilibrium in the cell

Diffusion (*top left*) is the passive movement of molecules from high to low concentration. Active transport (*left*) requires energy. The molecules are "bound" by chemicals on one side of the wall and taken through to the other side. Osmosis (*above*) is the movement of water through the semi-permeable cell wall, from a weaker to a stronger solution.

Simple and complex cells

The epithelial cells of the tongue are shown (*above*) magnified 500 times. They are among the simplest cells in the body, with a tough, resistant cell membrane surrounding the cytoplasm and nucleus. They are layered together and firmly attached to the underlying tissue. Nutrition reaches them by diffusion through the interstitial spaces and their only function is to replace themselves as a firm, waterproof layer.

Other simple cells are found in the skin or the nails and hair, where substances like keratin are produced.

Within the body more complicated cellular processes occur involving chemical excretions; for example digestive enzymes from the pancreas and hydrochloric acid in the stomach. Every minute 300 million cells in the body die and are replaced immediately by division of the living cells, so that the number of cells in the adult body remains constant throughout life.

Hormone control

The circulation brings hormones, such as thyroxine, and the corticosteroids from the adrenal gland; these help regulate the speed of the cell's activity. Some hormones have a specific effect on one type of cell, for example thyroid-stimulating hormone from the anterior pituitary gland, which affects only the thyroid gland. Other hormones affect one type of activity in all cells, such as insulin from the pancreas, affecting carbohydrate metabolism.

Chromosomes

Every human cell contains 46 chromosomes (*right*), except the sperm and ova, which have only 23. The chromosomes are compressed into the nucleus as chromatin. The sex chromosomes called X (red) and Y (blue) are circled. A combination of the longer X and shorter Y denotes male characteristics, an XX pair those of the female.

Each chromosome consists of a double strand of deoxyribonucleic acid (DNA) molecules. Each huge molecule of DNA is a gene. The combination of about 500 genes on each of the 46 chromosomes carries sufficient instructions for all the activities within the body, the way it is built and all the inherited instincts, from the colour and size of our eyes to the speed of our reactions.

Forty-six chromosomes

Photo: Ken Moreman

Cell division—mitosis

In mitosis (*right*) chromosomes become shorter (1) and the nuclear envelope breaks (2), releasing the chromosomes, which duplicate and attach themselves to a cytoplasmic network (3). They are then drawn apart (4-6) to form two new cells with re-formed nuclear envelopes (7).

Mitosis—cell reproduction

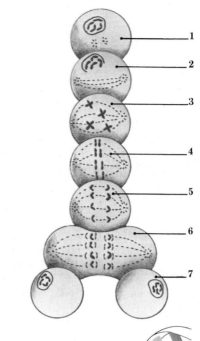

DNA—the code of life

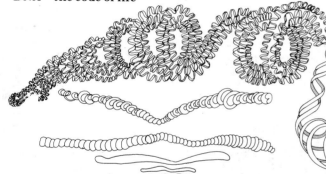

2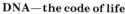

DNA—the code of life

DNA is the basic unit of control of human life. It is a highly complex substance formed from a chain (*above*) of chemical units—nucleotides. Each nucleotide is made from a sugar, phosphate and one of four different kinds of nitrogen-containing compounds—bases. The sugar and phosphate form the side of a ladder and the bases link in the middle to form the rungs of this spiralling ladder (1), the so-called double helix. These are the genes; it is the varying chemical composition of the DNA of each that causes the different RNA production.

When a cell is about to divide, the DNA molecules reproduce themselves by doubling and splitting lengthwise. When the nuclear envelope breaks, the two identical halves attach themselves to a cytoplasmic network that divides them.

DNA works by a temporary break in the double helix and the production of RNA, a sugar–phosphate–base complex similar but slightly different to the DNA. The RNA leaves the nucleus and acts as a guide for protein formation by the ribosomes on the endoplasmic reticulum. This new protein may be used elsewhere in the body or during the duplication of the DNA in mitosis.

The gradual magnification of a simple gene (2) to the final molecular structure of DNA illustrates its complexity. The model (*right*) is only a tiny portion of the whole molecule. The different bases are shown in varying colours.

Model of DNA

Photo: Courtesy Prof. M. H. F. Wilkins, King's College

How the Cell Works 3

Pituitary gland Most important endocrine gland, situated under the brain. It has two parts, the anterior lobe and posterior lobe.

Recessive Weak.

Ribonucleic acid (RNA) Chemical produced by DNA that controls the cell's activity by reproducing proteins on the ribosome.

Ribosome Area on the endoplasmic reticulum where the RNA reproduces proteins.

Sex-linked Genes that are found on the sex chromosomes.

Sperm Male gamete.

Thyroid gland Endocrine gland in the neck that produces thyroxine.

Thyroid-stimulating hormone Hormone produced by the anterior pituitary lobe that controls activity of the thyroid gland.

Thyroxine Hormone produced by the thyroid gland that helps control metabolism of the body.

Zygote New cell, containing 46 chromosomes, that is formed by the joining together of two gametes, each containing 23 chromatids.

Heredity

Normally a cell divides and the two new cells contain the same number of identical chromosomes. This is mitotic division. Human cells contain 23 pairs of chromosomes.

Reproductive cells—gametes—are formed by the splitting of the chromosome pairs to produce 23 chromatids in each cell. This is meiotic division. If this were the only thing to happen all the children of any one couple would be identical.

Before meiosis occurs an apparently normal mitotic cell division takes place, but the chromosomes overlap and seem to "stick" to each other. When they are pulled apart by the cytoplasmic network, the chromosomes break and change sections with each other, thus altering the genes. This is known as "crossing over" (*lower right*).

As each gamete is different, this alteration of genes gives each person a unique individuality. There are many features that will be nearly identical in close relatives and this makes members of the family recognizable as relations. Identical twins occur when the zygote—fertilized ovum—splits mitotically, producing two identical cells.

When the chromatids meet to form new pairs—alleles—certain physical features are opposite each other and, like the chromosome, become paired. There are probably 50,000 pairs of genes in each zygote carrying all the future instructions for the body. Many of these pairs will carry identical instructions—known as homozygous—but if they are different—heterozygous—one gene will have "stronger" instructions than the other. This is known as the dominant gene. The weaker one is called the recessive gene. Some genes are not always dominant or recessive; one may only *tend* to dominate.

The majority of genes do not vary and are homozygous. Most people have the same number of features and grow in a very similar manner. If the genes are badly damaged it is unlikely that the zygote will develop at all. Minor genetic changes may produce abnormalities, usually harmful but occasionally beneficial.

Meiosis—gamete formation

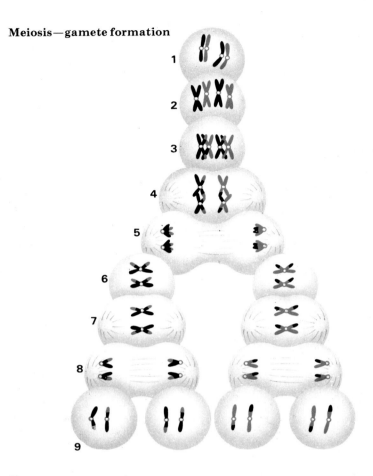

Chromosomes crossing over

Gamete formation

Gamete cells are produced to transmit genetic information (*above*) from parents to their children.

The chromosomes appear in the nucleus (1).

The DNA splits and the chromosomes are duplicated as the nuclear envelope breaks (2).

Similar chromosomes overlap and adhere to each other—"crossing over"—exchanging sections (3).

The cytoplasmic network pulls the chromosomes apart to line them up in the centre (4).

The chromosomes are then pulled to opposite poles of the cell (5).

The cell membrane breaks and two new cells are formed, each containing 46 non-identical chromosomes (6).

Meiosis can now start (7).

The two chromosome pairs split apart and each half, a chromatid, is pulled by the cytoplasmic network to the side of the cell (8).

The cell membrane breaks and the two new pairs of cells are formed, each containing 23 chromatids. This is a gamete. The joining of two gametes causes pairing of the chromatids and formation of a new cell, the zygote, with 46 chromosomes (9). Occasionally more or less than 46 chromosomes are formed, which usually causes a severe disorder.

Hemophilia—European royal families

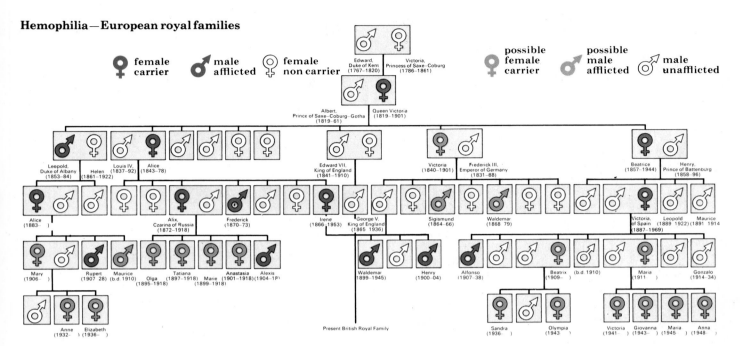

Legend:
- ♀ female carrier
- ♂ male afflicted
- ⊙♀ female non carrier
- ♀ possible female carrier
- ♂ possible male afflicted
- ⊙♂ male unafflicted

Edward, Duke of Kent (1767–1820) — Victoria, Princess of Saxe–Coburg (1786–1861)

Albert, Prince of Saxe–Coburg–Gotha (1819–61) — Queen Victoria (1819–1901)

Leopold, Duke of Albany (1853–84) · Helen (1861–1922) · Louis IV. (1837–92) · Alice (1843–78) · Edward VII, King of England (1841–1910) · Victoria (1840–1901) · Frederick III, Emperor of Germany (1831–88) · Beatrice (1857–1944) · Henry, Prince of Battenberg (1858–96)

Alice (1883–) · Alix, Czarina of Russia (1872–1918) · Frederick (1870–73) · Irene (1866–1953) · George V. King of England (1865–1936) · Sigismund (1864–66) · Waldemar (1868–79) · Victoria, of Spain (1887–1969) · Leopold (1889–1922) · Maurice (1891–1914)

Mary (1906–) · Rupert (1907–28) · Maurice (b.d. 1910) · Olga (1895–1918) · Tatiana (1897–1918) · Marie (1899–1918) · Anastasia (1901–1918) · Alexis (1904–18) · Waldemar 1899–1945 · Henry (1900–04) · Alfonso (1907–38) · Beatrix (b.d. 1910) · Maria (1911) · Gonzalo (1914–34)

Anne (1932–) · Elizabeth (1936–)

Present British Royal Family

Sandra (1936–) · Olympia (1943–) · Victoria (1941–) · Giovanna (1943–) · Maria (1945–) · Anna (1948–)

Hereditary defects

Many of an individual's characteristics are due to the balance in his genetic inheritance. Most characteristics are formed by more than one gene, such as eye colour; others are produced by the dominance of one gene only, such as that in blood groups and sub-groups. Rare conditions are due to the combination of two recessive genes; this can be seen in albinism—lack of the pigment melanin—when both parents may be "carriers" without any sign of the condition.

Sex-linked conditions are due to the gene being absent in the smaller male allele of the pair of sex chromosomes, so that the gene on the female half asserts itself. Hemophilia, an inherited deficiency in a blood-clotting factor, is such a condition. From the family tree of the royal families of Europe (above), it seems probable that Queen Victoria spontaneously produced the gene for hemophilia. Colour-blindness is another sex-linked condition and is ten times more common in males than females. As a colour-blind man may marry a "carrier" some of the daughters may also be colour-blind.

Sex of an individual

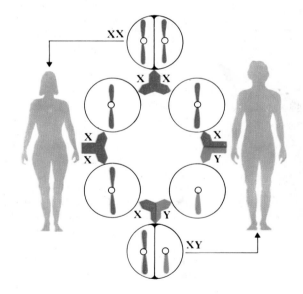

The sex of an individual

An individual's sex is determined by a particular pair of chromosomes (above). If they are homozygous a female is produced and the two alleles are of equal length with the same number of genes. These are called XX. If the pair is heterozygous, XY, a male is produced. The Y allele is smaller and contains fewer genes than the X half of the pair. Thus the unpaired genes act in a dominant manner.

Eye colour

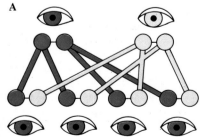

A

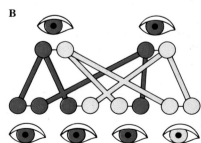

B

Eye colour

There is more than one gene for eye colour (above), but brown is dominant over blue. Two people, one with two genes for brown eyes, the other with two genes for blue eyes (A), will have children who all have brown eyes. However, if two brown-eyed parents (B) carry the heterozygous recessive blue gene they will have one blue-eyed child for every three brown-eyed children.

Cross-references

Eye 67
Pregnancy 84–5

Skeleton 1

Basic facts

The skeleton consists of about 206 bones divided into two broad groups, the axial and appendicular skeletons. It has three functions: It supplies support; it protects the internal organs; and, by using muscles, it gives movement. The axial skeleton, consisting of the skull, spine and rib cage, supplies the basic structure on to which the limbs, the appendicular skeleton, are joined via the pelvic and shoulder girdles.

Appendicular skeleton—upper limb

The shoulder girdle consists of two bones, the clavicle and scapula. The clavicle acts as a strut with one end fixed against the manubrium sterni and the outer end holding the scapula away from the thorax. The glenoid cavity is for the joint with the humerus.

The arm consists of long bones and a highly mobile hand. The humerus joins the radius and ulna at the elbow and the mobility of these two bones allows pronation and supination of the forearm and hand.

Appendicular skeleton—lower limb

The pelvis is much stronger than the shoulder girdle as it has to support the full weight of the body. Each innominate is formed from three bones: The wing-shaped ilium; the pubis in front; and the ischium behind. These are fused together. In front the innominate bones articulate at the symphysis pubis and join the massive sacrum behind. All these three bones form the acetabulum for the articulation of the femur. The lower and upper limbs are similar but with different functions. The long femur has its head extended to the side by the neck to articulate in the acetabulum. Large tubercles— the greater and lesser trochanters— are formed for muscle insertion. The lower end of the femur articulates the tibia, with the patella in front; the fibula merely supplies support for the muscles and part of the ankle joint. The seven tarsals and five metatarsals support the body weight; the fourteen phalanges are much smaller in the hand as they have little active function.

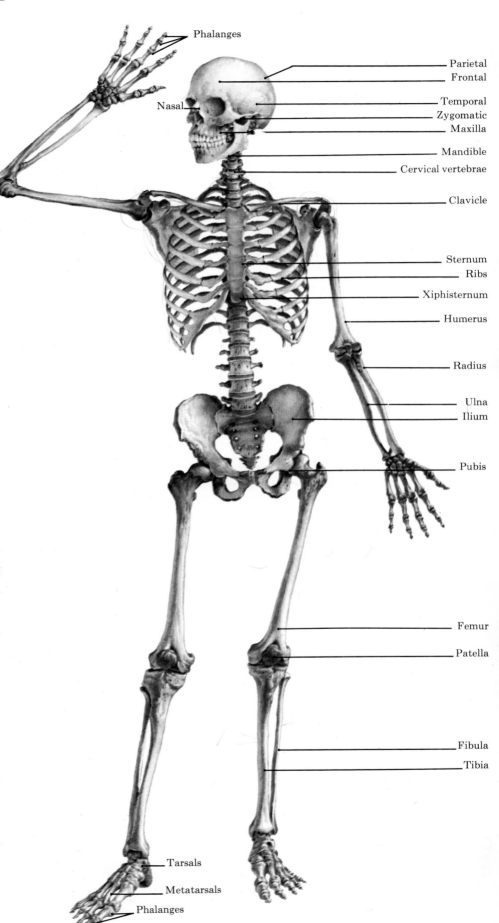

Phalanges

Nasal

Parietal
Frontal
Temporal
Zygomatic
Maxilla
Mandible
Cervical vertebrae
Clavicle

Sternum
Ribs
Xiphisternum
Humerus

Radius

Ulna
Ilium

Pubis

Femur
Patella

Fibula
Tibia

Tarsals

Metatarsals

Phalanges

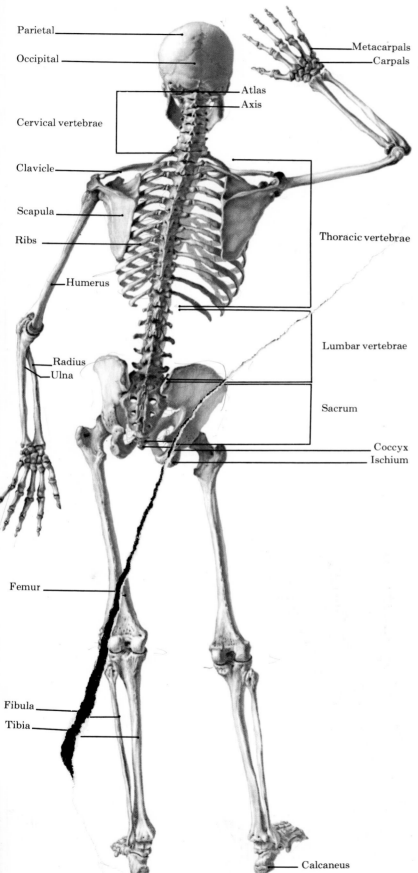

Parietal

Occipital

Cervical vertebrae

Clavicle

Scapula

Ribs

Humerus

Radius
Ulna

Femur

Fibula
Tibia

Metacarpals
Carpals

Atlas
Axis

Thoracic vertebrae

Lumbar vertebrae

Sacrum

Coccyx
Ischium

Calcaneus

Axial skeleton—spine
The spine consists of 7 cervical, 12 thoracic and 5 lumbar vertebrae. The 5 sacral and 4 vertebrae of the coccyx are fused together to make solid bone.

Axial skeleton—thorax
The thorax consists of 12 pairs of ribs, articulating with the thoracic vertebrae, 10 pairs joined with cartilaginous processes to the sternum in front leaving the two lowest pairs "floating". The sternum consists of the manubrium, at the top, and the small xiphi-sternum at the lower end.

Some people have an extra vertebra or an extra rib.

Bone structure
All bones have an outer, compact, dense layer and inner, spongy, cancellous centre. This makes them strong and light. They also act as storage for calcium and phosphorus and in many bones the cancellous centre is replaced by a medulla containing the marrow—blood-forming cells—or, in the case of sinuses, an air-containing space.

The articular surfaces of bone are covered with cartilage to supply a smooth surface for the joint. The overall surface of most bones is irregular and grooved by blood vessels or nerves. Bones are surrounded by tough, fibrous periosteum into which muscles and ligaments are inserted. Their traction causes ridges, tubercles or crests to occur by periosteal reaction—new bone formation. There is no nerve supply to bone, but blood vessels enter through the nutrient canal to reach the cancellous centre.

Bone growth
Growth takes place in all bones, but is more obviously apparent in the long bones. An infant's long bones have bony ends—epiphyses—separated from the shaft by the cartilagenous diaphyseal line of new bone formation. By the age of 25 (earlier in women) all these diaphyseal lines have ceased growing and have fused with the bone. Short and irregular bones, like carpal bones and the maxilla, calcify from the centre. Flat bones start as a membranous sheet that forms bone and gradually thickens to a centre of cancellous bone.

Definitions

Acetabulum Cup-shaped cavity on the side of the pelvis in which the head of the femur articulates.

Calcify Deposition of calcium salts in a tissue.

Cancellous "Honey-comb" bone structure.

Cartilage Special, tough, firm form of connective tissue, the degree of hardening depending on the number of collagen fibres.

Cervical Concerning the neck.

Collagen Tough connective tissue.

Conchae Three shell-shaped structures on the lateral wall of each nasal passage, 2 of cartilage and 1 of bone.

Cranium Bones of the skull that protect the brain.

Diaphysis Cartilagenous strip separating the epiphysis from the shaft to the long bone which is concerned with growth.

Epiphysis Piece of bone separated from the shaft of the long bone by the diaphysis.

Fontanelle Unossified area, between bones, in the skull of an infant.

Foramen Hole in a bone.

Glenoid cavity Area of articulation between the humerus and scapula.

Haversian system Tubular system of which bone is formed. A central canal surrounded by calcified tissue.

Interstitial fluid Body fluid outside the circulation that bathes the cells.

Lumbar Concerned with the lower back.

Meninges Membranes that surround the brain and spinal cord.

Skeleton 2

Common diseases

Achondroplasia Dominant inherited absence of the diaphyseal area leading to loss of growth of the long bone and dwarfism.

Acromegaly Characteristic overgrowth of the bones of the skull and ends of the long bones occurring from excessive secretion of growth hormone after normal growth has ceased.

Dwarfism This may be due to a variety of causes, e.g. achondroplasia.

Fracture Comminuted—broken into more than two pieces. Complicated—other structures, e.g. nerves, are damaged. Compound—in which the skin or surface membrane is broken. Greenstick—only one side of the bone is broken, the other side being bent. Impacted—the two fragments are driven together causing internal rigidity. Simple—the bone is broken in one place only without involvement of other structures.

Bone formation

All bones are formed from cartilage, with the exception of the clavicle and some of the skull, which ossify directly from membrane.

Compact bone is surrounded by periosteum and made of long columns of calcified tissue. These columns are known as Haversian systems (*right*), each with a central canal containing an artery, vein and lymph vessels, and filled with interstitial fluid. Osteoblasts—bone-forming cells—lay down concentric rings—lamellae—of mineral salts, mainly calcium and phosphorus, in the framework of the cartilage. These layers have collagen fibres running in different directions to each other to give greater strength. Long bones are also formed from the periosteum.

In cancellous bone the Haversian systems are less densely packed and the collagen fibres are in a honeycomb network pattern. This gives lightness in contrast to the strength from compact bone.

Bone formation

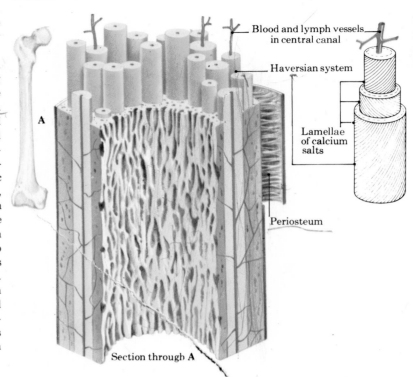

Blood and lymph vessels in central canal

Haversian system

Lamellae of calcium salts

Periosteum

Section through **A**

The hand

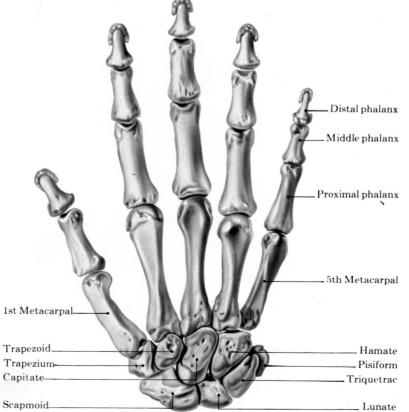

Distal phalanx

Middle phalanx

Proximal phalanx

5th Metacarpal

1st Metacarpal

Trapezoid

Trapezium

Capitate

Scapmoid

Hamate

Pisiform

Triquetrae

Lunate

The foot and the hand

Although they have very different uses the foot and the hand are similar in structure. The foot bears the full weight of the body and acts as a spring when walking and running. The massive calcaneum thrusts the weight on to the ground at the back; this is transferred to the front of the foot by the higher medial and smaller lateral longitudinal arches to reach the bases of the first and fifth metatarsal bones. There is a smaller transverse arch between these two points that turns the foot into a triangle of arches held together by the ligaments, tendons and muscles of the foot.

The hand (*left*) is far more delicate, with long, easily mobile fingers, each containing 3 phalanges. A thumb, of 2 phalanges, has the unique ability to reach across the palm, making grasping 5 metacarpals, thus making the 8 carpal possible. The mobility of the fingers bones in the wrist, with an instrument and thumb, has produced the eyes associated with the skilled and brain, able to carry out skilled and accurate movements in *a way that* would be impossible *with the foot.*

Exploded view of the skull

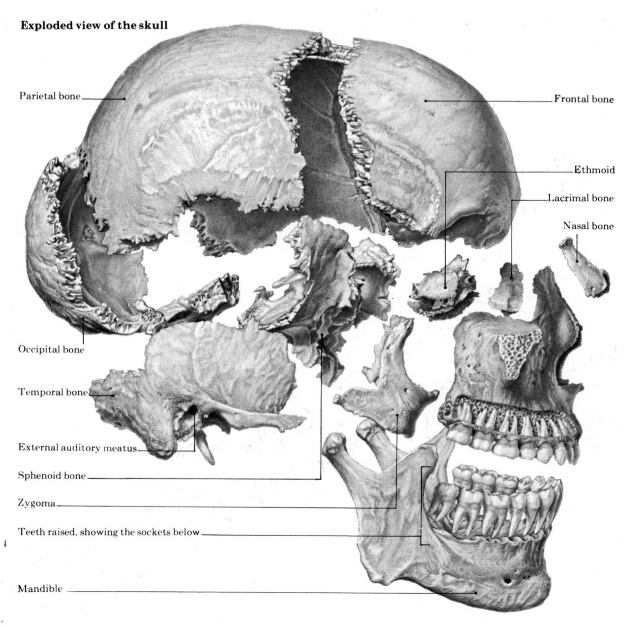

Parietal bone

Frontal bone

Ethmoid

Lacrimal bone

Nasal bone

Occipital bone

Temporal bone

External auditory meatus

Sphenoid bone

Zygoma

Teeth raised, showing the sockets below

Mandible

The skull

The skull (*above*) is both the shell of protective bone for the brain, eyes and organs of hearing, and the structure of the face.

The eight bones of the cranium—occipital; 2 temporal; 2 parietal; frontal; sphenoid; ethmoid—are fused together. This rigid structure is pierced by foramina for the cranial nerves and blood vessels. It is marked internally by the meningeal arteries and externally by ridges at muscle insertions. The occipital bone con-tains the foramen magnum and articulates with the atlas. The sphenoid bone contains a cup-like cavity to protect the pituitary gland. The three bones of the middle ear are deep inside each temporal bone. In infants, before the bones fuse, fontanelles are found anteriorly between the parietal and frontal bones and posteriorly between the parietal and occipital bones.

The face is composed of fourteen bones: 2 maxillae; 2 nasal; 2 zygoma; 2 lacrimal; the mandible and (not shown in the picture) 2 palatine; 2 inferior turbinate bones (the superior and middle nasal conchae are carti-lagenous) and vomer (the inferior and posterior parts of the nasal septum).

The mandible, like the maxilla, contains sockets for the 32 teeth, which are embedded in fibrous tissue. The nerves from the organ of smell—cranial nerves—penetrate the nasal part of the ethmoid bone. The fron-tal, maxillary, ethmoid and mastoid process of the temporal bones con-tain air sinuses to lighten the skull.

Cross-references

Whole Body 13
Joints 24–5
Blood 35

Joints

Definitions

Acetabulum Cup-shaped cavity on the side of the pelvis.

Atlas First cervical vertebra.

Axis Second cervical vertebra.

Bursa Small sac filled with fluid.

Cartilage Special tough, firm form of connective tissue.

Cruciate ligament One of two internal ligaments found in the centre of the knee joint.

Disc Fibro-cartilaginous structure between vertebrae.

Fontanelle Area of the infant's skull that is not covered with bone. It gradually closes up during development.

Ligamentum flavum Ligament running inside the neural canal between the arches of the vertebrae.

Ligamentum teres Ligament from the head of the femur to the centre of the acetabulum.

Neural arch Arch of bone posterior to the vertebrae.

Neural canal Channel within the neural arches.

Neural foramina Gap between the neural arches through which the spinal nerves pass to reach the body.

Nucleus pulposus Soft jelly-like centre of the invertebral disc.

Quadriceps femoris Major thigh muscle that causes extension of the knee joint.

Semilunar cartilage One of a pair of C-shaped cartilages on either side of the knee.

Suture Joint line between the bones of the skull.

Symphysis pubis Joint between the two pubic bones.

Basic facts

A joint is the meeting point between bones; it usually allows a controlled amount of movement. Some joints have to be very strong, while others have to be very mobile. It is not possible for a joint to be both strong and mobile. Joints can be classified according to their mobility:

Mobile joints

Because they have to be able to withstand the friction of movement these joints (*below and right*) are complex. The surfaces of the bone are covered with smooth cartilage. The joint edge has a strong fibrous capsule surrounding a sac of synovial membrane between the bone ends. This membrane secretes a small amount of lubricant fluid that allows frictionless movement and the articular cartilage keeps the two bones apart. The joint has stabilizing ligaments which bind and strengthen it.

Slightly mobile joints

Some joints require a small amount of movement, but still have to be very strong. This can be seen between the vertebrae of the spine and the symphysis pubis. In these joints there is a thick pad of fibro-cartilage between the bones held in place by a strong fibrous ligament. This pad acts as a shock absorber.

Immobile joints

In infancy many of the skull bones are not joined together and in two places, the anterior and posterior fontanelles, there are gaps between them. As the infant grows these open joints join to form the rigid skull. The irregular, serrated edges of bone are bound together by tough, fibrous tissue at the suture lines.

Mobile joints

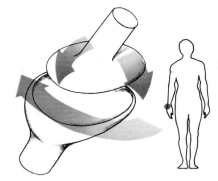

Ellipsoid joint Allows circular and bending movement but no rotation, e.g. between finger and palm.

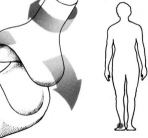

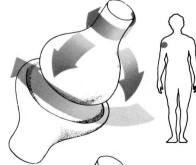

Saddle joint Allows movement in two directions, but without rotation, e.g. ankle, and thumb joints between wrist bones and thumb.

Ball-and-socket joint A joint freely moving in all directions, e.g. shoulder and hip joints.

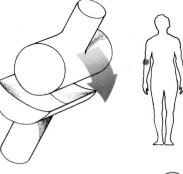

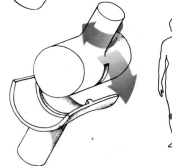

Hinge joint Allows extension and flexion, e.g. elbow and finger joints.

Condylar joint This is similar to a hinge joint, but with slight rotation to allow the joint to "lock" into an extended position, e.g. the knee joint.

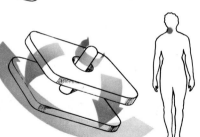

Plane joint A flat surface allows the bones to slide on each other, but they are restricted by ligaments to a small range, e.g. tarsal bones of foot and between ribs and thoracic vertebrae.

Pivot joint Allows rotation, but no other movement, e.g. between the atlas and axis.

The spine

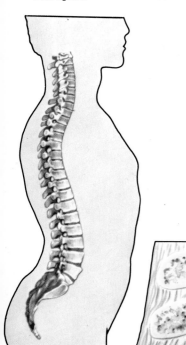

The spine—discs

The main joints between the vertebrae of the spine (*left*) are slightly mobile. The vertebral surface (*below*) is covered with hyaline cartilage (1) and the intervening space is filled with a thick ring of fibro-cartilage (2) with a centre of soft, almost gelatinous, tissue—the nucleus pulposus. The joints are held together by the anterior (3) and posterior (4) longitudinal ligaments, the ligamentum flavum (5) in the neural canal (6), the interspinous (7) and supraspinous (8) ligaments.

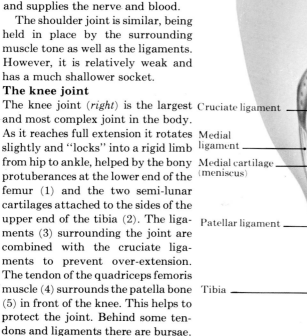

Neural foramina
Anterior longitudinal ligament
Posterior longitudinal ligament
Hyaline cartilage
Intervertebral disc, fibro-cartilage
Neural canal
Interspinous ligament
Ligamentum flavum
Supraspinous ligament

The spine—movement

The spinal vertebrae also have joints between their other articulating surfaces, on the neural arches, and with the ribs in the thoracic region. These have a synovial membrane and are surrounded by ligaments. This allows a much greater degree of movement. The joints between the atlas and occiput and between the axis and atlas do not have discs but rely on synovial membranes to give freedom of movement.

The intervertebral discs act as "shock absorbers" for the skull and brain. The movement between individual vertebrae, with the exception of the axis and atlas, is small, but the overall combined effect is considerable. Most of the flexion and extension is in the cervical and lumbar regions, while bending to the side is principally in the thoracic area. Twisting involves the whole vertebral column. Local movement allows the nerves to slide out of the neural canal through the neural foramina (9), without damage.

Ball-and-socket joints

In the hip joint the head of the femur fits into the acetabulum and is held in place by a rim of cartilage. Inside the joint the strong ligamentum teres holds the femur to the pelvis and supplies the nerve and blood.

The shoulder joint is similar, being held in place by the surrounding muscle tone as well as the ligaments. However, it is relatively weak and has a much shallower socket.

The knee joint

The knee joint (*right*) is the largest and most complex joint in the body. As it reaches full extension it rotates slightly and "locks" into a rigid limb from hip to ankle, helped by the bony protuberances at the lower end of the femur (1) and the two semi-lunar cartilages attached to the sides of the upper end of the tibia (2). The ligaments (3) surrounding the joint are combined with the cruciate ligaments to prevent over-extension. The tendon of the quadriceps femoris muscle (4) surrounds the patella bone (5) in front of the knee. This helps to protect the joint. Behind some tendons and ligaments there are bursae.

The knee joint

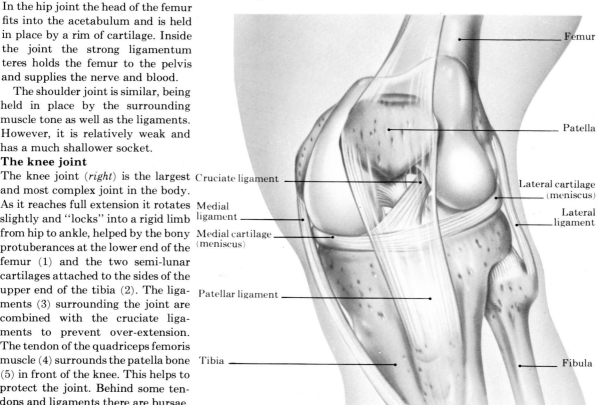

Femur
Patella
Cruciate ligament
Medial ligament
Medial cartilage (meniscus)
Lateral cartilage (meniscus)
Lateral ligament
Patellar ligament
Tibia
Fibula

Cross-references

Whole Body 13
Skeleton 20–3
Muscles 27

Muscles 1

Definitions

Acetylcholine
Chemical produced in the transmission of nerve impulses to the muscles by the autonomic nervous system.

Adrenaline Hormone produced by the medulla of the adrenal gland. It augments the action of the sympathetic nervous system.

Antagonist In this sense one of a pair of muscles whose balance of action is such that an antagonist gradually relaxes as the other muscle contracts.

Cramp Involuntary severe spasm of a muscle.

Fascia Thin layer of connective tissue.

Fibril Small area of contractile tissue in a muscle. Protein molecules are the basic component.

Ganglion A swelling. A nerve ganglion is where nerves join to each other outside the central nervous system.

Lactic acid One of the substances produced when muscle activity takes place.

Muscle "belly" Fleshy part of the muscle.

Muscle "head" Beginning of the muscle belly at the site of origin. A muscle may have more than one head.

Muscle "insertion" Muscle belly contracts down to form a tendon, which is attached to either a bone or aponeurosis, called its insertion.

Muscle "origin" Usually a large area where the muscle is attached to the bone.

Muscle "tone" Constant slight contraction that keeps the fleshy part of a muscle firm.

Noradrenaline One of the chemicals produced by the autonomic nervous system in nervous transmission to muscle.

Skeletal muscles

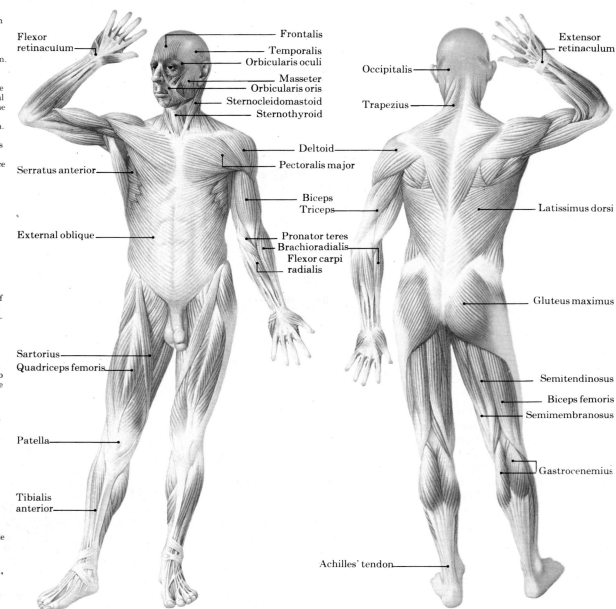

Flexor retinaculum — Frontalis — Temporalis — Orbicularis oculi — Masseter — Orbicularis oris — Sternocleidomastoid — Sternothyroid — Serratus anterior — Deltoid — Pectoralis major — Biceps — Triceps — Pronator teres — Brachioradialis — Flexor carpi radialis — External oblique — Sartorius — Quadriceps femoris — Patella — Tibialis anterior

Occipitalis — Trapezius — Extensor retinaculum — Latissimus dorsi — Gluteus maximus — Semitendinosus — Biceps femoris — Semimembranosus — Gastrocnemius — Achilles' tendon

Basic facts

Muscles are contractile tissue, which can initiate or maintain movement in the body. Muscles comprise 35 to 45 per cent of the total body weight and there are over 650 skeletal muscles (those directly under the skin are shown above) controlled by the nervous system. There are three kinds of muscle (shown top right): Skeletal, under voluntary control; cardiac, only found in the heart; smooth, as in the intestine.

Contraction

Muscles are composed of fibrils, which shorten when chemically activated. This occurs in response to nervous stimulus. The chemical that initiates this depends on the type of muscle, but the response is the same —a shortening of protein molecules. As soon as the nerve stimulus ceases, the fibril relaxes and the muscle lengthens. Muscle tone is maintained by some of the fibrils always being stimulated to contract.

Muscles—metabolism

When muscles contract energy is required and heat produced. This metabolism produces carbon dioxide, lactic acid, heat and water. The blood flow in to the muscles is increased to take away the metabolites, the heart rate increases and the heat has to be dissipated through the skin and by sweating. Excessive use leads to a local build-up of lactic acid as the oxygen supply is not sufficient to metabolize it. This can cause cramp.

Types of muscles

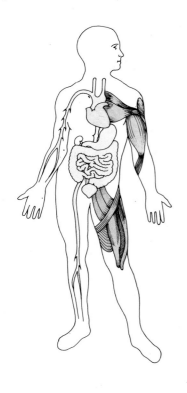

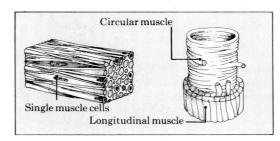

Circular muscle

Single muscle cells

Longitudinal muscle

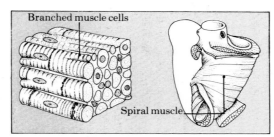

Branched muscle cells

Spiral muscle

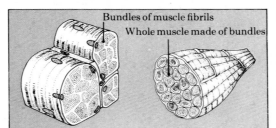

Bundles of muscle fibrils
Whole muscle made of bundles

Smooth muscle
Made up of long cells arranged in bundles, each with a single nucleus. The intestine has a circular inner layer producing constriction, and an outer longitudinal layer to produce wave-like peristalsis. Controlled by the autonomic system, smooth muscle contraction is involuntary.

Cardiac muscle
Made up of branched cells with many nuclei that interweave and form thick, spiral bands around the ventricles of the heart. These have a slow, rhythmical contraction that is increased by the cardiac pacemaker. Cardiac muscle is also controlled by the autonomic nervous system.

Skeletal muscle
Made up of long cells, some as much as 20 centimetres in length, composed of minute myofibrils and nuclei, that form bundles under the control of the voluntary nervous system. Skeletal muscle can contract and relax rapidly, but it is easily fatigued.

Skeletal muscles—anatomy
These may be massive, like the gluteus maximus in the buttock, or minute, like the stapedius muscle inside the middle ear. Most muscles join one bone with another and have their "origin" on one immobile bone and "insertion" on the other, the one that is moved. The origin is usually on a moderately large area of bone and, with some muscles, may be from more than one area, as on the two "heads" for the biceps. The main body of the muscle is called "the belly". The insertion is usually a tendon attached to a small area of bone, but may be an aponeurosis attached to other structures, as found in the back. Long tendons passing over other structures, for example at the wrist, are usually surrounded by sheaths of synovial membrane to reduce friction. In these areas the tendons are held down by the retinaculum.

Short tendons insert directly into the periosteum of the bone and often produce a small bony nodule on the surface of the underlying bone.

Skeletal muscle—movement
Long muscles, like the sartorius, contract further than stronger muscles like the deltoid. This is because the arrangement of fibres is parallel, while those in the deltoid are arranged in a fan shape of small bundles inserting into the tendon.

Muscles are usually arranged in pairs (right), so that as one contracts (1) the other, the antagonist (2), slowly relaxes to give a smooth, controlled movement. Muscles are gently pulling against each other all the time. This is muscle tone.

When a muscle contracts and moves a joint it can only do so if the insertion is held rigid. The brain controls skeletal muscle movement and co-ordinates movement using information relayed from the muscle itself, and from the eye and the organ of balance in the ear.

Muscles requiring skilled movements, as in the hand, have one nerve to a few fibres, but in those that are used for strength, such as the gluteus maximus, one nerve will supply a large number of fibres.

Joint movement

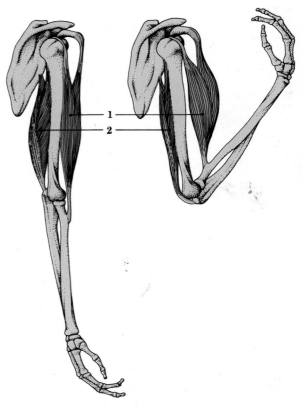

1
2

Muscles 2

Common diseases

Fibrositis Local areas of oedema (swelling) cause stiffness and discomfort in muscle groups. This may occur after unusual exercise.

Injuries Muscle injuries are common after strains and may be accompanied by the inability to use the muscles normally.

Muscular dystrophy Group of similar congenital muscle disorders leading to increasing weakness and paralysis.

Myasthenia gravis Disorder disease of the transmission of impulses between the nerve ending and the muscle, leading to fatigue and weakness when the muscles are used

Autonomic nervous system

Many muscles in the body are controlled by the autonomic nervous system (*right*), composed of sympathetic nerves (red) and parasympathetic nerves (blue). These involuntary muscles are of two kinds: Cardiac muscle, which is only found in the heart; and smooth muscle.

Smooth muscles are found all over the body—surrounding blood vessels regulating the blood flow; in the bronchioles controlling the entry of air to the alveoli; in the iris of the eye, dilating and contracting the size of the pupil; and attached to the hairs in the skin, making them "stand on end". Larger amounts are found making up the walls of the bladder and uterus and their appendages—ureters and Fallopian tubes. They are also found from about half-way down the esophagus.

Release of urine

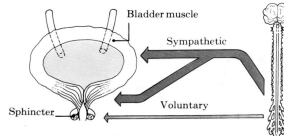

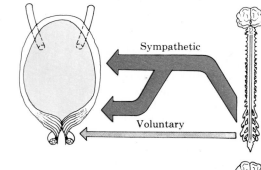

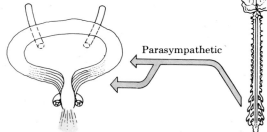

Autonomic nervous system

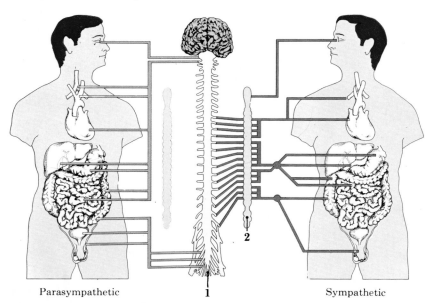

Parasympathetic   1   Sympathetic

Release of urine
The bladder fills with urine. The sympathetic nerves to the bladder muscle make it relax and the sphincter contract. Voluntary nerves also control the sphincter so that urine can be passed at convenient times.

The urine pressure increases so the sympathetic nerve impulses intensify, keeping the sphincter firmly closed and the bladder relaxed. In the normal active male the bladder holds approximately 300 millilitres of urine.

When the bladder is full the sympathetic nerve impulses cease and the parasympathetic nerves cause the bladder to contract and the sphincter to open. Only controlled voluntary impulses can prevent the urine from being passed.

Involuntary control

The intestinal wall is surrounded by a network of nerve fibres from both the sympathetic and parasympathetic nervous systems. The sympathetic nerves release noradrenaline to produce relaxation of the muscle walls and contraction of the sphincters. The parasympathetic nerves release acetylcholine to produce the opposite effect—contraction of the muscle walls and relaxation of the sphincters. The stretch receptors in the smooth muscle help in the control of muscle tone.

The balanced control of these two systems produces peristalsis. The autonomic nervous system (*above*) is ultimately under the control of the brain's hypothalamus. The sympathetic nerves leave the spinal cord (1) to reach ganglia (2)—nervous relay stations—which interconnect and lie on the muscles near the spine before passing on to the body. The sympathetic nervous system prepares the body for action—dilated pupils, cold skin, raised blood pressure and pulse rate, and blood diverted from a relaxed intestine to the skeletal muscles. The parasympathetic nerves run directly from the spinal cord. The parasympathetic nervous system prepares the body for a relaxed state.

Peristalsis

The alimentary canal extends from the pharynx (1) to the rectum (2). When the food enters it, circular muscle fibres contract (3) behind it and relax in front (4), while the outer, longitudinal, fibres (5) squeeze it along in a wave-like fashion. This is known as peristalsis and is the manner in which food is taken through the alimentary canal.

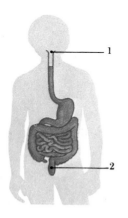

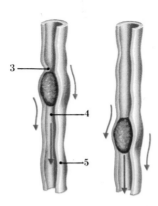

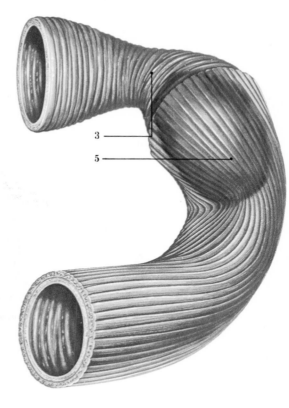

Peristalsis

When food enters the mouth it is under voluntary control and can be chewed and even spat out. Once it has entered the esophagus it is no longer under conscious control and is moved down the alimentary canal by peristalsis (above). Peristalsis is a complex, reflex wave of contraction and relaxation. The mass of food is called a "bolus". The circular and longitudinal smooth muscle fibres behind the bolus contract and this tends to push it onwards. This tendency would be frustrated if the fibres over the front part of the bolus and those immediately in front of it did not relax and create a space into which it could move. As it moves forwards the wave of constricting muscle extends forward to squeeze the bolus from behind.

Peristaltic waves occur all the time and move the intestinal contents along the alimentary canal. Sometimes peristalsis moves backwards and forwards mixing the contents—particularly in the stomach—before it is released in small amounts at a time into the duodenum.

Sphincters

Tight areas of circular muscles—sphincters—close off one part of the intestine from another. They only open when it is necessary for food to pass through. There are six sphincters: Esophageal sphincter, where the esophagus joins the stomach; pyloric sphincter, between the stomach and duodenum; the small sphincter, between the terminal ileum and colon; the anal sphincter, at the end of the alimentary canal, which is also under voluntary control; sphincter of Oddi, where the common bile and pancreatic ducts join the duodenum; and the sphincter in the bladder.

When food is eaten the empty stomach stimulates a giant peristaltic wave. Several of these waves pass along the entire length of the alimentary canal, moving the contents onwards. The contents of the colon are moved to the rectum, which is normally empty. The distension of the rectum produces the urge to defecate. This is why so many people do so after the first meal of the day. Antiperistaltic waves are peristalsis in reverse and will produce vomiting.

Cardiac muscle

Cardiac muscle is arranged in thick interweaving circles around the ventricles. It has an inbuilt rhythm of contraction and relaxation which occurs about 80 times a minute. Special, electrically sensitive muscle fibres in the "pace-maker" respond to the autonomic nervous system and produce a faster, regular rhythm adjusted to the body's needs.

The blood pressure is produced by the strength of the heart beat and is reflexly controlled by sensitive nerve endings responding to the pressure in the carotid arteries.

Involuntary muscle and hormones

Adrenaline, released from the adrenal glands, augments the action of the sympathetic nervous system; to a lesser extent thyroxin, from the thyroid gland, has a similar effect.

Muscle and heat production

Heat is produced by peristalsis and the maintenance of muscle tone. Cooling of the body causes rapid, involuntary contractions of skeletal muscle—shivering. It is the contractions that make extra heat.

Myotonia Group of congenital muscle disorders of increased tone and stiffness.

Myositis Inflammation of the muscle. This will commonly occur with any acute infectious illness and is a transitory affair. Rarely, there is true infection of the muscles.

Polymyalgia rheumatica Inflammation of the muscle arteries in the elderly causing stiffness and pain of muscles and joints.

Rheumatoid arthritis Although this is a disease mainly of the joints, the muscles are also involved and will become weaker.

Rupture of Achilles' tendon Any tendon may rupture and one of the commonest is the largest in the body, the Achilles' tendon, which may spontaneously rupture after minimal strain. This is probably due to a disorder of the collagen connective tissue of which it is made.

Synovitis Inflammation of the synovial membranes covering the tendons and joints.

Tendinitis Inflammation, usually due to excessive use, of the tendons.

Heart and Circulation 1

Definitions

Aorta Largest and most important artery in the body.

Aortic valve Valve between the left ventricle and aorta.

Arteriole Very small artery, with muscular coat, which divides to form capillaries.

Artery Blood vessel carrying blood from the heart.

Atrioventricular node Small area of electrically sensitive muscle that lies between the atria and ventricles at the top of the interventricular septum and is joined to the Bundle of His.

Atrioventricular valve Valve between the atrium and ventricle, mitral or tricuspid.

Atrium One of the two thin-walled chambers of the heart.

Baroreceptors Nerve endings sensitive to changes in blood pressure, present in the aorta and carotid arteries.

Basilar artery Artery that continues beneath the midbrain to join the two vertebral arteries.

Bundle of His Specialized cardiac muscle conducting electrical impulses from the atrioventricular node to the ventricles.

Capillary Smallest of blood vessels.

Cardiac centre Area of nervous control of the heart that is situated in the medulla oblongata.

Carotid artery Pair of arteries running from the aorta to the brain and to the outside of the head.

Carotid body Contains chemoreceptors near carotid sinus.

Carotid sinus Slight swelling of the beginning of the internal carotid artery containing baroreceptors.

Position

The heart (*right*) lies in the thorax behind the sternum and in front of the descending aorta and esophagus. It rests on the central ligament of the diaphragm muscle. On either side are the two lungs. Above are the main blood vessels and the bifurcation of the trachea into the two main bronchi.

Basic facts

The heart weighs about 300 grammes and is approximately the size of a grapefruit. It has two atria, two ventricles and four valves. It receives blood from the two venae cavae and four pulmonary veins and expels it into the aorta and pulmonary artery. It pumps 9,000 litres of blood a day at a rate varying between 60 and 160 beats a minute.

Anatomy

The heart is surrounded by the fibrous pericardium containing the serous pericardial sac holding a small amount of fluid that allows frictionless movement. It consists of two pairs of chambers—atrium and ventricle—that act as separate pumps. The right side pumps deoxygenated blood through the lungs—the pulmonary circulation. The left side pumps oxygenated blood from the lungs—the systemic circulation.

Atria and ventricles

The blood from the superior and inferior venae cavae enters the right atrium. The four pulmonary veins take blood to the left atrium.

The atrioventricular valves, tricuspid in the right and mitral in the left, have special muscles—the papillary muscles—and fine tendons—chordae tendineae—attached to the edges of the valve cusps to prevent them rupturing back into atria during ventricular systole.

The left ventricle has thicker muscle than the right to sustain a higher systemic blood pressure. The ventricles are closed by the aortic and pulmonary valves.

The heart is lined with endocardium, and divided into its two halves by the interatrial and interventricular septa.

The remnants of the fetal circulation can still be seen in the fibrous band of the ductus arteriosus.

The heart

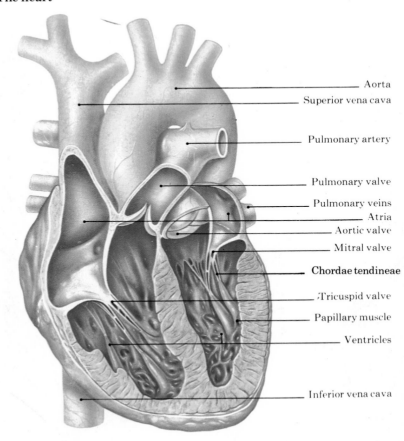

- Aorta
- Superior vena cava
- Pulmonary artery
- Pulmonary valve
- Pulmonary veins
- Atria
- Aortic valve
- Mitral valve
- **Chordae tendineae**
- Tricuspid valve
- Papillary muscle
- Ventricles
- Inferior vena cava

Electrical system

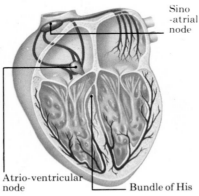

- Sino-atrial node
- Atrio-ventricular node
- Bundle of His

Coronary circulation

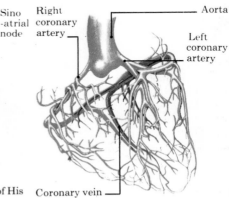

- Right coronary artery
- Aorta
- Left coronary artery
- Coronary vein

The electrical system

To keep the heart (*above*) beating the sino-atrial node, "pace-maker", in the right atrium sends impulses through the two atria, causing atrial systole. It then stimulates the atrioventricular node to pass rapidly down the Bundle of His to cause ventricular systole.

Coronary circulation

The myocardium has its own blood supply (*above*) from the left and right coronary arteries—the first branches of the aorta—supplying different areas of muscle with a few anastomoses between them. The venous blood is drained into the right atrium by the coronary veins.

The blood vessels

The arteries carry blood away from the heart and the veins return it. Both arteries and veins (*right*) are made in the same way with four layers: A protective fibrous coat; a middle layer of smooth muscle and elastic tissue, which is thickest in the largest arteries; a thin layer of connective tissue; and a smooth layer of cells—endothelium.

Arteries

The blood in the arteries is under high pressure, greater in the systemic than the pulmonary circulation, so the arterial walls are stronger, with thicker muscular and elastic layers. This allows the artery to expand with the surge of pressure at each heart beat and thus smooth out the blood flow. This can be felt as a "pulse" in arteries near the surface.

The larger arteries divide into smaller ones and finally into arterioles, where the blood flow can be controlled by autonomic nerves supplying the smooth muscle. This allows adjustments of blood flow to various areas, such as increased flow to the intestine after a meal.

Veins and venous blood flow

The blood reaches the venous system through the minute capillary vessels. It is through the capillary wall that oxygen and carbon dioxide, food and metabolites are exchanged with the interstitial fluid. Most of the interstitial fluid returns to the venous system, but some is collected by the lymphatic vessels.

The returning venous blood moves slowly due to low pressure and the veins can collapse or expand to accommodate variations in blood flow (*right*). Movement relies on the surrounding muscles, which contract (1) and compress the vein. Pulsation of adjacent arteries (2) has a regular pumping effect.

Semilunar valves (3) are found at regular intervals throughout the larger veins and these allow the blood to move only in one direction. They are commoner in the legs.

The veins frequently anastomose with each other so that the blood flow can alter direction if there is any constriction or pressure from movement of muscles or ligaments.

The blood vessels

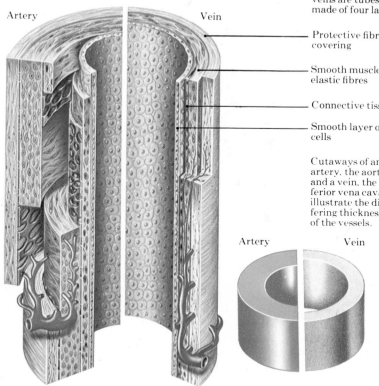

Artery   Vein

Both arteries and veins are tubes made of four layers:

Protective fibrous covering

Smooth muscle and elastic fibres

Connective tissue

Smooth layer of cells

Cutaways of an artery, the aorta, and a vein, the inferior vena cava, illustrate the differing thicknesses of the vessels.

Artery   Vein

Venous blood flow

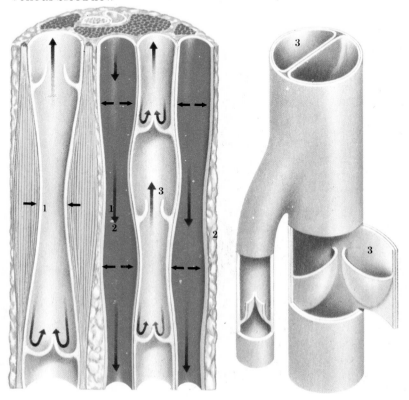

Cerebral artery One of three pairs—anterior, middle and posterior arteries—supplying the cerebral hemisphere.

Chemoreceptor Nerve ending sensitive to chemical changes.

Chordae tendineae Fibrous strands from the papillary muscles to the edges of the atrioventricular valves.

Coronary "Crown" of two blood vessels supplying the blood to the heart muscle.

Diastole Period of relaxation of the heart muscle.

Ductus arteriosus Fetal blood vessel that connects the aorta and pulmonary artery and which closes at birth.

Endocardium Layer of smooth cells lining the internal surface of the heart.

Endothelium Smooth layer of cells lining the internal surface of the blood vessels.

Foramen ovale Inter-atrial opening that closes at birth.

Iliac artery Aorta divides into two iliac arteries.

Lymphatic vessel Thin-walled vessel that collects interstitial fluid and returns it, as lymph, to the venous system located in the neck.

Medulla oblongata Part of the brain stem.

Mesenteric Supplying the intestine.

Mitral valve Valve between left atrium and ventricle.

Myocardium Muscle of the heart.

Papillary muscle Ventricular muscle to which the chordae tendineae are attached.

Pericardium Sac surrounding the heart.

Portal system Vein beginning and ending in capillaries.

Heart and Circulation 2

Pulmonary Refers to structures associated with the lungs.

Pulmonary valve Valve between the right ventricle and pulmonary artery.

Pulse Wave of pressure with each heart beat that can be felt in the peripheral arteries near the surface.

Semilunar valve Valves found in veins. So-called because each cusp of the valve looks like a half-moon.

Septum Thin structure dividing two parts, e.g. the inter-atrial septum.

Sino-atrial node Electrically sensitive tissue in the right atrial muscle initiating heart beat.

Sinus arrhythmia Alteration in heart rate associated with respiration.

Sphygmoman-ometer Instrument for recording blood pressure.

Systemic Circulation to the whole body.

Systole Period of heart muscle contraction when blood is pumped into the arteries.

Tachycardia Rapid heart beat.

Tricuspid valve Valve between the right atrium and ventricle. It has three cusps.

Valve Structure allowing fluid to flow in only one direction.

Vein Blood vessel carrying blood back to the heart.

Vena cava One of the two principal veins returning blood from the systemic circulation to the right atrium.

Ventricle Strong muscular chamber of the heart.

Vertebral artery One of a pair of arteries that runs within the cervical vertebrae to form the basilar artery at the base of the brain.

The two circulations

The systemic circulation carries oxygenated blood and nutrition and returns with metabolites that have to be filtered and excreted. The pulmonary circulation quickly exchanges carbon dioxide and oxygen through the alveolar surface in the lungs.

The circulatory system

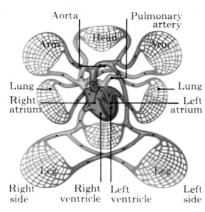

The circulatory system

Blood is pumped into large and then smaller arteries (*above*) before reaching the arterioles and finally the capillaries, each about 1 millimetre long and one-hundredth of that in diameter. It is here that the exchange of gases takes place and interstitial fluid is formed before the blood returns into the venous system. The systemic and pulmonary circulations are based on the same system with lymphatic vessels collecting excess interstitial fluid.

Kidney and liver circulations

Twenty per cent of the cardiac output passes through the kidneys. This blood is at high pressure to maintain adequate filtration of metabolites into the renal tubules.

The liver has a small hepatic artery and large hepatic veins due to the portal venous system. This starts as capillaries in the intestine, from the upper end of the stomach to the anus, where anastomoses with the systemic venous system occur. The blood flows in the mesenteric veins to join the splenic vein to form the hepatic portal vein and enter the liver, where it again forms capillaries to enter the hepatic sinusoids.

Anatomy of the circulation

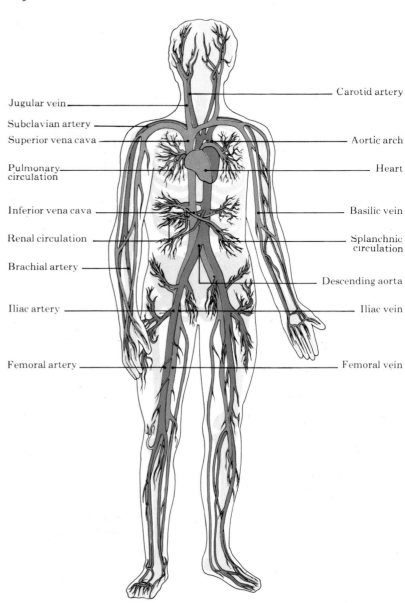

Anatomy of the circulation

The blood is carried in a 96,500 kilometre network of blood vessels (*above*). The aorta leaves the aortic valve where its first two branches, the coronary arteries, subdivide into brachial and carotid arteries. The aorta then runs in front of the spine and behind the esophagus to the abdomen, with small and large branches to the vertebrae,

diaphragm and intercostal muscles. In the abdomen it divides just above the pelvis into the common iliac arches to the legs.

There are four main branches: Gastric to the stomach; splanchnic to the intestine; renal to the kidneys; and splenic to the spleen. The venous blood returns to the superior and inferior venae cavae and thus to the right atrium.

The heart in health

A baby's heart rate is about 130 beats a minute; this falls to around 70 in the resting adult. A trained athlete can raise the rate to about 200 with great exertion. The heart rate is controlled from the cardiac centre in the medulla; parasympathetic fibres in the vagus slow the rate and sympathetic nerves and adrenaline increase it in response to stressful situations, such as emotional disturbance and fear.

Chemoreceptors, in the carotid body, are sensitive to low oxygen levels and cause a tachycardia. Baroreceptors in the carotid sinus reflexly control the blood pressure.

The cerebral circulation

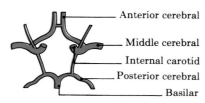

— Anterior cerebral
— Middle cerebral
— Internal carotid
— Posterior cerebral
— Basilar

The cerebral circulation

The two vertebral arteries join to make the basilar artery, which communicates with the two internal carotid arteries to make a circle of vessels (above) from which the anterior, middle and posterior cerebral arteries supply the cerebral hemispheres. This ensures an adequate blood supply at all times.

Blood pressure

Blood pressure has two measurement points: Systolic, the highest pressure reached; and diastolic, the lowest, when the aortic valve is closed and the left ventricle is in diastole. The elasticity of the artery wall allows expansion and then maintains arterial blood flow, during diastole, by contraction. A loss of elasticity with ageing is associated with higher systolic pressures. Blood pressure is measured with a mercury column in a sphygmomanometer. Relaxed, healthy adults have a blood pressure of around 120 millimetres systolic, and 70 millimetres diastolic, written as 120/70. Pressures over 150/90 are hypertensive.

Blood pressure rises slightly on standing as there is a nervous contraction of blood vessels in the legs.

A heart beat

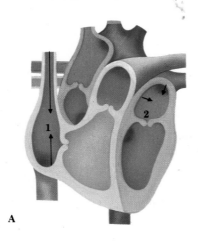

A

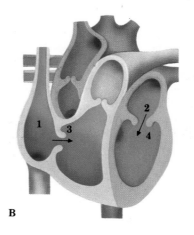

B

A heart beat

In atrial diastole (**A**) blood flows from the inferior and superior venae cavae into the right atrium (1) and from the four pulmonary veins into the left atrium (2). The flow is increased during inspiration as the negative intrathoracic pressure "sucks" blood into the heart as well as air into the lungs. This results in sinus arrhythmia.

When ventricular systole ceases (**B**) the intraventricular pressure drops and the two atrioventricular valves—tricuspid (3) and mitral (4)—float open and blood starts to flow from the atria (1, 2) into the ventricles. The sino-atrial node then initiates atrial systole and the blood is pumped past the valves, which are fully open, into the dilated ventricles.

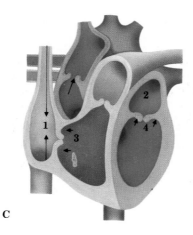

C

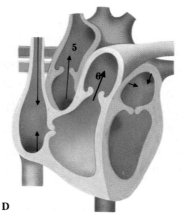

D

Atrial systole ceases (**C**) when the electrical impulse reaches the atrioventricular node and passes down the Bundle of His to start ventricular systole. Once atrioventricular valves (3, 4) snap closed, the chordae tendineae and papillary muscles prevent them bursting back into the atria. Venous blood can again flow into the atria (1, 2), during atrial diastole and ventricular systole.

The rapidly increasing ventricular pressure (**D**) throws open the aortic (5) and pulmonary (6) valves with blood streaming into the systemic and pulmonary circulations. The elasticity of the arterial walls causes the valves (5, 6) to snap closed at the end of ventricular systole. The "snap" of the closing heart valves can be heard as "lub-dub" through the chest wall.

Common diseases

Endocarditis
Inflammation of the heart lining, usually the valves, e.g. bacterial endocarditis.

Myocarditis
Inflammation of the heart muscle, e.g. rheumatic myocarditis.

Pericarditis
Inflammation of the pericardium, e.g. tuberculous pericarditis.

Rheumatic fever
Raised temperature and inflammation of connective tissues—occurring mostly among children.

Septal defect
"Hole in the heart".

Diseases of blood vessels

Arteries—Arteriosclerosis Narrowing of the vessels leading to blockage and death of the area supplied, e.g. coronary or cerebral thrombosis.

Hypertension
Raised blood pressure, usually for no obvious reason.

Veins—Varicose
Congenital absence of valves leading to distortion of veins by the pressure.

Electrical irregularities

Atrial fibrillation
Irregular, rapid heart beat due to disorder of the sino-atrial node.

Heart block
Failure of the electrical impulse to reach the ventricles, which then contract at their "natural" rhythm of 40 beats every minute.

Cross-references

Muscle 26, 27, 29
Respiration 38–43

Blood 1

Definitions

Adenoid   Pad of lymphoid tissue on the posterior wall of the posterior nasal cavity.

Albumen   Simple plasma protein.

Antibody   Specific substance produced by the body in an attempt to dispose of a foreign substance.

Antigen   Any substance introduced in the body that causes the formation of an antibody.

Basophil   Blood cell that is stained by basic staining dye.

Blood group   Classification of different types of blood by their antigen characteristics.

Carbaminohemoglobin   Hemoglobin combined with carbon dioxide.

Carbonic anhydrase   Enzyme that causes water to combine with carbon dioxide to form carbonic acid.

Eosinophil   Blood cell stained red.

Fibrin   Protein that causes strands in clotting.

Fibrinogen   Soluble precursor of fibrin in the circulation.

Folic acid   Member of the vitamin B group.

Gamma globulin   Particular group of plasma protein associated with immunity.

Globulin   Major group of plasma proteins.

Hemoglobin   Complex protein that gives the colour to red cells and is essential for the carriage of oxygen. It contains the element iron.

Hemolysis   Breakdown of red blood corpuscles releasing hemoglobin.

Kupffer's cells   Modified macrophage cells of the reticulo-endothelial system in the liver's sinusoid.

Blood formation

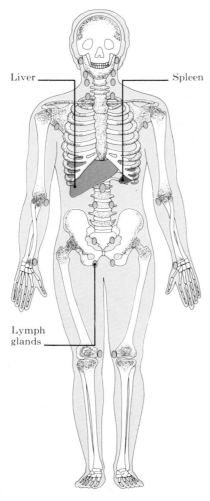

Liver

Spleen

Lymph glands

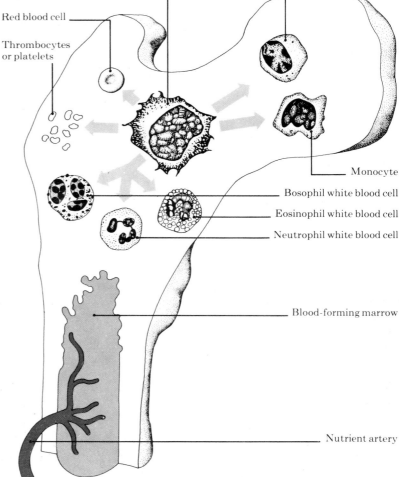

Stem cell

Red blood cell

Thrombocytes or platelets

Lymphocyte

Monocyte

Bosophil white blood cell

Eosinophil white blood cell

Neutrophil white blood cell

Blood-forming marrow

Nutrient artery

Blood

Blood is the fluid that is pumped round the body in the circulatory system. It is made up of many components and has many functions. A baby has about 250 millilitres, a man over 5 litres and a woman rather less, about 4 litres of blood. It consists of a yellow fluid—the plasma—in which red and white blood corpuscles and platelets are suspended. The capillaries allow fluid to escape from the blood. The cells and large proteins are left in the vessel and this fluid can now become the interstitial fluid. This will either return to the capillary or join the lymphatic system. In the brain it helps form the cerebro-spinal fluid and in the eyes the intraocular fluid. Blood amounts to about one-third of the total interstitial fluid.

Blood formation

Red and white blood cells (*above*) are formed in the bone marrow—the dotted areas of bone. In addition, white cells are produced by the liver, spleen and lymph nodes. The stem cells in the bone marrow produce the red and white blood corpuscles (magnified 1,000 times). The red corpuscles enter blood vessels through the nutrient circulation. The white corpuscles—the polymorphonucleocytes, basophils, neutrophils and eosinophils—have segmented nuclei. Monocytes and lymphocytes have single nuclei. Thrombocytes—platelets—are also formed from the stem cell system by fragmentation and are involved in blood clotting.

In infants, all the bones make red blood cells, and before birth the liver and spleen also produce them.

The plasma

The plasma occupies 55 per cent of the blood volume. It is 90 per cent water, 7 per cent proteins and lipoproteins with the remaining 3 per cent made up of small molecules—salts, glucose, vitamins, hormones, urea, amino-acids and dissolved carbon dioxide.

The main proteins are albumen, globulins and fibrinogen. They help to maintain the osmotic pressure to hold fluid in the circulation and counteract the blood pressure that forces it out of the capillaries. The proteins also repair the body, the gamma-globulins defend it against infection, and fibrinogen may break down to form fibrin when clotting occurs. Serum is plasma without fibrinogen and is seen as the thin fluid oozing from grazed skin.

Constituents of the blood

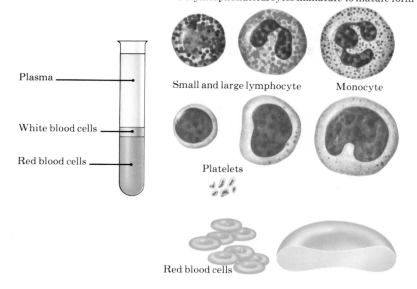

Polymorphonuclearcytes immature to mature form

Small and large lymphocyte   Monocyte

Plasma

White blood cells

Red blood cells

Platelets

Red blood cells

White blood corpuscles

There are normally about 8,000 white corpuscles (*left*) in every cubic millilitre of blood. There are two main varieties: Polymorphonuclear leucocyte (polymorphs) and lymphocyte. Polymorphs are so called because of their nuclei. Formed in the bone marrow stem cells, they make up two-thirds of white cells. The immature cell has one nucleus, but as it matures a string of up to five nuclei forms. They survive less than a week. There are three different kinds of polymorphs, distinguished by their microscopic staining characters: Eosinophils, basophils and neutrophils.

Polymorphs act by phagocytosis; they have two functions: Destruction of invading bacteria; and removal of dead and damaged tissue. If the polymorphs are killed pus is formed. The eosinophils react in allergic states.

Lymphocytes are formed in the lymph glands, spleen and thymus. They help with the immunity of the body. The monocytes are larger cells, distinct from lymphocytes, and function mainly by phagocytosis.

Platelets (thrombocytes) are tiny particles formed in the bone marrow; there are about 250,000 per cubic millimetre lasting less than a week. They are concerned with clotting.

Lipo-protein Substance formed from fat and protein.

Lymph Fluid collected by the lymphatic vessels from the interstitial spaces.

Lymphatic system Vessels that carry lymph and return it to the venous system.

Lymphocyte White blood cell with clear cytoplasm and a single nucleus formed in lymphoid tissue.

Lymphoid tissue Specialized tissue that produces lymphocytes, filters out foreign substances and helps in developing immunity.

Macrophage Fixed tissue cell that is involved in the phagocytosis of bacteria and foreign substances.

Marrow Blood-forming tissue in the centre of bone.

Monocyte Circulating phagocytic cell.

Neutrophil Body cell that does not stain with basic or eosinophilic dye.

Pus Collection of dead white blood cells.

Red blood corpuscle Hemoglobin-carrying cell without a nucleus.

Red blood corpuscles

Each red corpuscle (*above*) measures about 7.5 microns (thousandths of a millimetre) in diameter and is biconcave in shape. It contains hemoglobin, which gives it a red colour. There are 5 to 6 million red corpuscles per cubic millimetre of blood.

Red corpuscle production requires vitamins of the B complex—principally vitamin B_{12}—and folic acid; vitamin C; and, as well as proteins, the elements iron, copper and cobalt. The corpuscles are formed from the marrow stem cells, but less than 1 per cent of the circulating red cells have nuclei as the nucleus is removed when it reaches maturity

The red corpuscle survives about 120 days and the damaged and old cells are removed by the reticulo endothelial system in the spleen and liver. The red corpuscles have only one function: The transport of oxygen by combining with hemoglobin. Healthy men have about 15 grammes of hemoglobin in 100 millilitres of blood and women 13 grammes.

Blood clotting

Tissue factor

Fibrinogen

Platelet

Plasma factor

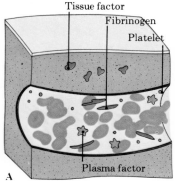 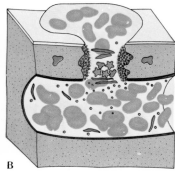 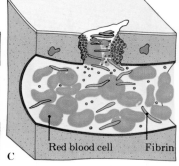 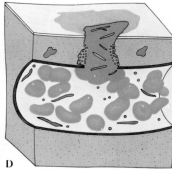

Red blood cell   Fibrin

A   B   C   D

Blood clotting

Blood clotting prevents excessive blood loss from a wound. Normally, circulating blood contains red cells, platelets, plasma, clotting factors and fibrinogen. Tissue-clotting factors lie trapped within cells surrounding each blood vessel (**A**). When damage occurs blood escapes from the broken vessel. Platelets congregate at the site and help plug the wound. Tissue-clotting factors are released (**B**). The reaction of the platelets with plasma and tissue-clotting factors converts the soluble fibrinogen into insoluble threads of fibrin. The fibrin forms a mesh across the break (**C**). Platelets and blood cells become trapped in the mesh. The jelly-like mass shrinks and serum oozes out, leaving a clot (**D**).

Blood 2

Reticulo-endothelial system Cells that are found throughout the body that are concerned with phagocytosis of foreign material, bacteria and broken tissues, and also the development of immunity.

Rhesus factor Blood antigen found in 85 per cent of the population that is independent of the ABO blood groups.

Serum Liquid that is left after blood has clotted.

Stem cell Basic, primitive cell from which all blood cells and platelets are formed.

Thoracic duct Largest lymphatic vessel; it runs from the posterior part of the thorax and joins the venous system in the neck.

Thrombocyte *See* Platelet.

Thymus gland Gland situated behind the upper sternum that is concerned with the development of immunity, until late adolescence, by the production of a lymphoid-stimulating hormone.

Transfusion reaction Antigen–antibody reaction that occurs when an incompatible blood is transfused.

Urea Nitrogen-containing substance formed from ammonia in the liver.

White blood corpuscles Concerned with fighting infection.

Blood groups

Although the red cells in different people look the same they are, in fact, dissimilar. They can be divided into four main groups, A, B, AB, and O *(below)*. The surface of the cells in each group is different and will act as an antigen to plasma from another group, which carries the antibody. This causes the cells to stick to each other—agglutination. An individual with group A cells will carry the antibody B in his plasma, those in group AB do not carry either antibody, while those in group O have both antibodies but the cells do not have either antigen. Thus blood from any group can be transferred into those of group AB, as they do not carry antibodies; group AB individuals are known as "universal recipients". A group O recipient can only receive blood from another group O donor, but can give blood to anyone, and thus is known as a "universal donor". Antibodies in the plasma of the donor blood are quickly diluted by the recipient and the concentration is thus too low to cause agglutination. The two commonest groups in western Europe are groups A and O, each found in about 45 per cent of the population, group B in about 10 per cent and group AB in less than 5 per cent.

The antigen–antibody reaction not only causes agglutination but also hemolysis—the breakdown of the red blood cells releasing hemoglobin into the circulation. This is an "incompatible transfusion reaction" and can lead to fever, jaundice, kidney failure from blocking of the tubules with hemoglobin, and even, in some cases, death.

Rhesus factor

In addition to the four main blood groups there are many other minor ones: Rhesus, NNS, P, Kell Lewis, Duffy, Lutheran, to name a few. The most important is the Rhesus factor, named after the Rhesus monkey, in which it was first discovered. The presence or absence of this antigen makes the individual either Rhesus positive or Rhesus negative. Antibodies are not found in Rhesus negative people unless they have been transfused with Rhesus positive blood. About 15 per cent of the population is Rhesus negative. It is therefore possible to be group A Rhesus positive or negative and this is usually written A Rh +ve or A Rh −ve. A true universal donor has to be O Rh −ve. Blood from the recipient and donor are cross-matched before use to check that agglutination does not occur.

Pregnancy and the Rhesus factor

Particular problems arise in Rhesus negative pregnant mothers with Rhesus positive babies. In the last weeks of pregnancy small numbers of the baby's red cells escape through the placenta into the mother's circulation. This will not cause trouble in the first pregnancy but in subsequent pregnancies agglutination reactions may occur. The mother's Rhesus antibodies will invade the baby's circulation, destroying the red cells and leading to anemia; this necessitates an exchange transfusion for the baby at birth with Rhesus negative blood. Provided the possibility is recognized early enough anti-Rhesus immune globulin injections can be given to the mother, to remove Rhesus positive cells from her blood before she develops her own anti-Rhesus globulin antibody.

Blood: Oxygen and carbon dioxide

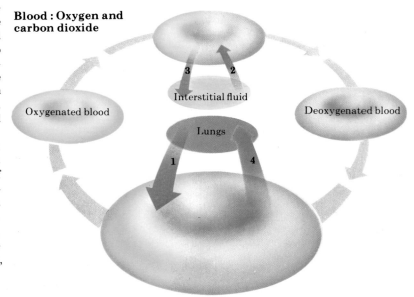

Oxygenated blood · Interstitial fluid · Deoxygenated blood · Lungs

Blood groups

Donor \ Receptor	A	B	AB	O
A				
B				
AB				
O				

Blood: Oxygen and carbon dioxide

Oxygen combines with hemoglobin (1) in the red blood corpuscles *(above)* to form oxyhemoglobin. It would require 75 times as much blood to carry the oxygen if it did not exist. Carbon dioxide (2). from the interstitial fluid, dissolves in the plasma and red cells, where the enzyme—carbonic anhydrase—forms carbonic acid. A little also combines with hemoglobin to form carbaminohemoglobin. The acidity of carbonic acid forces oxygen out of oxyhemoglobin (3) into the interstitial fluid with a formation of sodium and potassium bicarbonate. A reverse diffusion occurs in the capillaries of the lung alveoli with loss of carbon dioxide (4) and reoxygenation of the hemoglobin (1).

Immunity and resistance to infection

The body's first line of defence—the skin—is a barrier against infection; it is a waterproof layer with bactericidal sweat and sebaceous secretions. Internal surfaces are covered with a layer of mucus; the mucus is produced as a result of the cleansing effect of the ciliated cells lining the sinuses and bronchi and the antiseptic hydrochloric acid in the stomach. Peyer's patches in the small intestine engulf the remaining bacteria. The adenoids, tonsils and lingual pad of lymphoid tissue filter off bacteria in the nasopharynx. When infection enters the body there is an immediate response by the white blood cells, which invade the area and form pus if sufficient numbers are killed.

Throughout the body the reticuloendothelial system of macrophage cells will engulf foreign particles and bacteria. These cells are found principally within the bone marrow, lymphatic system and spleen. In the liver they line the sinusoids and are known as Kupffer's cells.

Circulating plasma globulins give protection against any infection, but this is greatly increased by successful resistance to an infection. This immunity follows the production of specific antibodies by the lymphoid tissue and reticulo-endothelial system. The thymus gland produces a hormone to stimulate immunity by the production of lymphoid tissue and lymphocytes. After adolescence it ceases to function.

Active, often permanent immunity is given by natural infection, injections of killed organisms, for example, diphtheria, or mildly infective living viruses, such as measles or, by mouth, poliomyelitis. Passive, temporary immunity is given by injections of gamma globulin prepared from a person who has been actively immunized, for example, human antitetanus globulin.

To achieve the maximum effect some vaccines have to be given in two or three doses, and a "booster" dose may be needed after a time to sustain immunity.

The lymphatic system

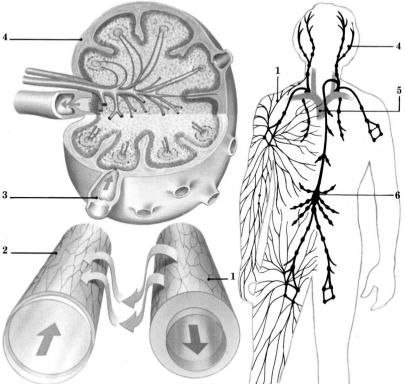

Lymphatic system

The lymphatic system (*above*) is a network of vessels that drains the interstitial fluid back into the blood circulation. The fluid passes out of the capillary (1) and either into the vein or into the smallest, thin-walled lymph vessel (2). These lymphatic vessels join together to form large channels and finally reach the thoracic duct running next to the descending aorta. This duct joins one of the main branches of the superior vena cava (5).

Valves (3) keep lymph flowing in one direction. Lymph glands—nodes (4)—are found throughout the body and at places where lymph vessels unite, in the groin, axillae and neck and adjacent to branches of the aorta and inferior vena cava (6). They have three main functions: To filter out and destroy foreign substances; to produce lymphatic cells; and to produce antibodies. Lymphoid tissue—specialized lymph glands—are the tonsils, adenoids and Peyer's patches in the ileum.

The spleen

The spleen (*left*), in the left upper abdomen, is a mass of lymphoid tissue. The white pulp produces lymphocytes and is scattered throughout the red pulp, in which macrophages filter out old or damaged cells, debris and bacteria. The spleen is also a store of iron from the broken red cells; and, like other lymphoid tissues, is an aid in the development of immunity. In adults, surgical removal of the spleen has no ill effects.

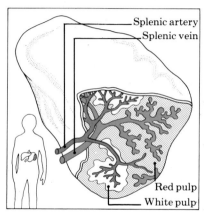

Splenic artery
Splenic vein
Red pulp
White pulp

Common diseases

Addisonian anemia Larger than usual red blood cells that are fewer in number due to a deficiency of vitamin B_{12} and associated with the destruction of the super-renal glands.

Anemia Lowered number of red blood cells in the circulation leading to a reduced ability to carry oxygen to the tissues. Aplastic anemia—due to failure of the bone marrow to produce red cells. Chronic disease, or hormonal deficiency, reduces the bone marrow's ability to produce red cells. Hemolytic anemia—unusually rapid destruction of red cells. Nutritional anemia—due to deficiency in necessary substances, such as vitamin B_{12}.

Coagulation defects These lead to spontaneous bleeding. This may be due to an absence of platelets.

Hemophilia Sex-linked coagulation defect due to the congenital absence of a necessary protein.

Leukemia Cancer of the white blood cells. This may be rapidly progressive—acute; or very slow—chronic.

Myeloma Type of malignant disease arising in the bone marrow and causing disorders of both the red and white blood corpuscles. It is a chronic disorder.

Thrombocytopenia Absence or severe reduction in the platelets, leading to spontaneous or post traumatic bleeding.

Cross-references

Heart 30–3
Respiration 38–43
Pregnancy 80–5

Respiration 1

Definitions

Adenoid Pad of lymphoid tissue on the posterior wall of the posterior nasal cavity.

Alveolar duct Final division of a bronchiole as it enters the alveolar sac.

Alveolar sac Collection of alveoli that are bunched together and supplied by one alveolar duct.

Alveolus Microscopic sac-like structure in which the exchange of gases in the lung takes place.

Bronchiole Small division between the small bronchi and the alveolar ducts. It is surrounded with contractile smooth muscle.

Bronchus Two main bronchi divide into smaller bronchi, which are the breathing tubes leading into the lungs. They have cartilaginous rings to give them rigidity.

Cilia Hair-like surface of the cells that line the bronchi and bronchioles. Their movements tend to sweep material upwards out of the lungs.

Clavicle Collar bone, from the upper end of the sternum to the shoulder.

Dead air space Air in the respiratory passages—trachea, bronchi and bronchioles—not involved in the gas exchange of respiration.

Diaphragm Wide, dome-shaped muscle that extends from the periphery at the ribs to a central tendon beneath the heart. It is pierced by the oesophagus and main blood vessels.

Dyspnoea Shortness of breath.

Epiglottis Small, flat, thin piece of cartilage attached to the larynx in the front and overlying the glottis.

Esophagus Muscular tube, lined with mucous membrane.

The lungs

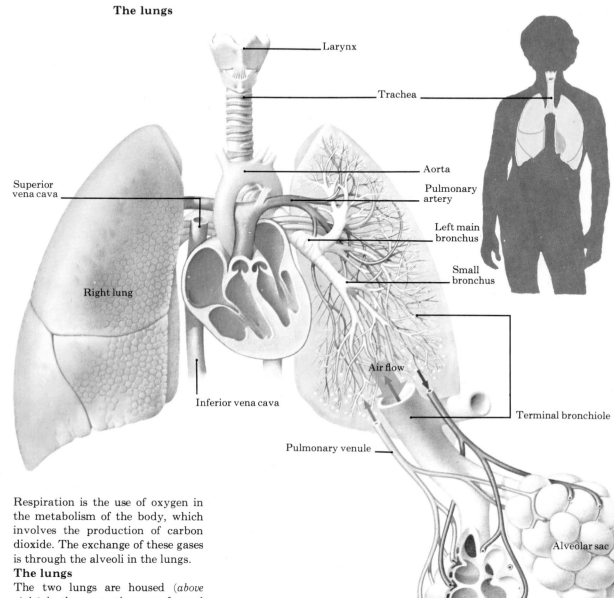

Respiration is the use of oxygen in the metabolism of the body, which involves the production of carbon dioxide. The exchange of these gases is through the alveoli in the lungs.

The lungs

The two lungs are housed (*above right*) in the protective cage formed by the ribs radiating from the spine at the back to the sternum at the front. The curved diaphragm muscle spans the base of this beehive-shaped cage and is pierced by the oesophagus, aorta and inferior vena cava.

The lungs (*above*) are separated by the heart, blood vessels and oesophagus. The top—apex—of each lung reaches behind the first rib to the level of the clavicle, with the base resting on the diaphragm.

The left lung, consisting of two lobes, is slightly smaller than the right, with three lobes.

Each lung is surrounded by pleura. This is a double layer of membrane lining the inside of the ribs and completely enveloping the lung; it contains some lubricating fluid.

Air enters the body through the nose and mouth to reach the back of the throat—pharynx—before passing through the voice-box—larynx—to reach the windpipe—trachea. The trachea is about 12 centimetres long. It is kept open by C-shaped cartilages, which, when the trachea divides, continue into the two main bronchi to the lungs.

The two main bronchi subdivide as smaller bronchi to the lobes of each lung. Each lobe is composed of segments supplied by one bronchus. Bronchioles subdivide into the alveolar ducts of the alveolar sacs containing the individual alveoli. The pulmonary blood vessels, lymph ducts and nerves enter the "root of the lung" with the main bronchi.

Breathing

The lungs (*below*) occupy most of the thoracic cavity (**A**). They are elastic structures and it is this elasticity that helps with the movements of breathing.

The main muscles for breathing are the diaphragm and the intercostal muscles, which run between the ribs. The accessory muscles that can sometimes be used are the shoulder, neck and abdominal muscles. They are used to help regain breath after running, by resting the arms on a table. This rigidity of the arms allows the arm muscles to move the chest wall instead of the chest wall acting as the firm structure for arm movement to aid in breathing.

During normal breathing the diaphragm does most of the work. It contracts, becoming flatter, and the rib cage expands (**B**). This increases the volume of the thorax and air is drawn down the trachea into the lungs—known as inspiration. Expiration takes place passively by the natural elasticity of the lung tissue and is effortless movement (**C**).

More forceful active expiration can be produced by using the intercostal and abdominal muscles. The intercostal muscles can help the diaphragm with inspiration.

The resting respiratory rate in a healthy adult is 12 times a minute; a baby breathes at twice this speed.

A normal breath is 500 millilitres of tidal air, but only 350 millilitres is fresh air, as the first air entering the lungs is 150 millilitres of dead air space—the volume of air in the nose, trachea and bronchi. The total amount of air—the vital capacity—that can be inhaled is increased by forced inspiration and expiration. There is always some air left in the lungs—the residual volume.

Vital capacity	4,000 ml
Tidal air	500 ml
Expiratory reserve	1,000 ml
Inspiratory reserve	2,500 ml
Residual volume	1,000 ml
Total lung capacity	5,000 ml

Breathing
A At rest   **B** Inspiration   **C** Expiration

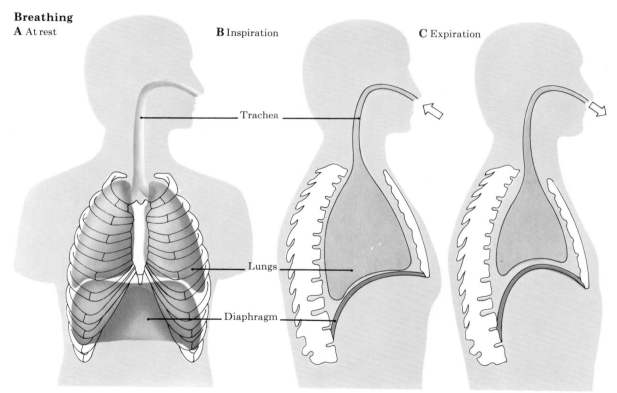

Trachea

Lungs

Diaphragm

The air in the nose, pharynx and the lungs

As air enters the nose larger dust particles are filtered out by the hairs in the nostrils. The air passes through the nose, where there is a large area of moist mucous membrane. This helps to humidify and warm it to body temperature. On cold, dry days the air is insufficiently warmed and moistened by the nose, so it will make the throat slightly sore—an effect often felt down to the trachea.

When air reaches the pharynx it passes through a region of lymphoid tissue in a ring around the back of the nose and throat. This consists of the adenoids on the posterior wall of the back of the nose, the two tonsils at the sides of the back of the mouth and an adenoid-like pad of tissue on the pharyngeal part of the tongue. This lymphoid tissue removes bacteria and viruses from inhaled air.

Air enters the lungs, where the smaller particles of dust are collected in the bronchioles. The lining of the bronchi and bronchioles is moist and covered with cilia that massage the mucus and debris upwards to the trachea and finally, over the larynx, into the oesophagus. Any dust particles in the alveoli are removed by the macrophage cells—the large white blood cells present in the walls of the alveolar sac.

Respiration 2

Anatomy of the alveolar sac

The anatomy of these minute grape-like alveoli is very delicate (*right*). They are held open by a framework of strands of fibrous connective tissue through which the terminal bronchioles, only 0·3 millimetres in diameter, pass and open into the individual alveolus.

The alveoli are lined by a thin layer of cells that coats the fibrous strands and the surrounding mesh of blood capillaries. The alveoli are kept moist by a thin film of water that is essential in the diffusion of oxygen and carbon dioxide in and out of the capillaries. Any dust, soot or bacteria that reaches them is ingested by the defensive macrophages in the moisture on the inner surface of the lining. Any excess fluid in the alveolar walls is drained away by the pulmonary lymphatic vessels.

The alveoli of an adult have a total surface area of 70 square metres; the whole breathing apparatus is designed to bring fresh air as close as possible to the blood.

Respiration and metabolism

The metabolism in the body requires oxygen to burn food, which is then converted into energy, carbon dioxide and waste substances, such as urea. This oxygen is obtained from the circulating oxyhemoglobin in the red blood corpuscles. There is only a small amount of oxygen dissolved in the plasma. The carbon dioxide is entirely dissolved in the plasma. Exercise or any fever will increase the breakdown of food, The increase in the production of carbon dioxide creates the need for more oxygen. This leads to an increased respiratory rate.

Some diseases, such as diabetes, may cause an increase of acid production, not carbon dioxide. This acid acts as a stimulant to cause an increased respiratory rate. The body is unable to differentiate between the acidity of carbon dioxide and acid due to other causes.

Diseases that increase the metabolic rate, such as over-activity of the thyroid gland—hyperthyroidism—contrast with the physical and respiratory sluggishness of an underactive thyroid—myxedema.

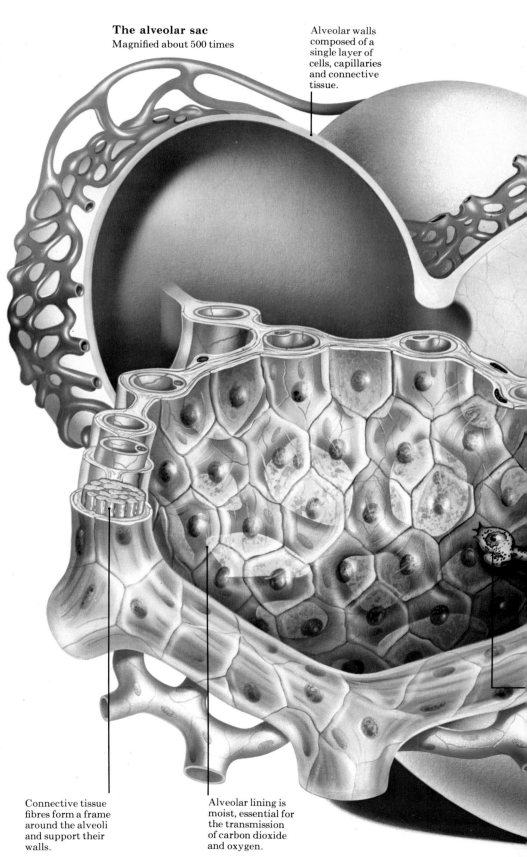

The alveolar sac
Magnified about 500 times

Alveolar walls composed of a single layer of cells, capillaries and connective tissue.

Connective tissue fibres form a frame around the alveoli and support their walls.

Alveolar lining is moist, essential for the transmission of carbon dioxide and oxygen.

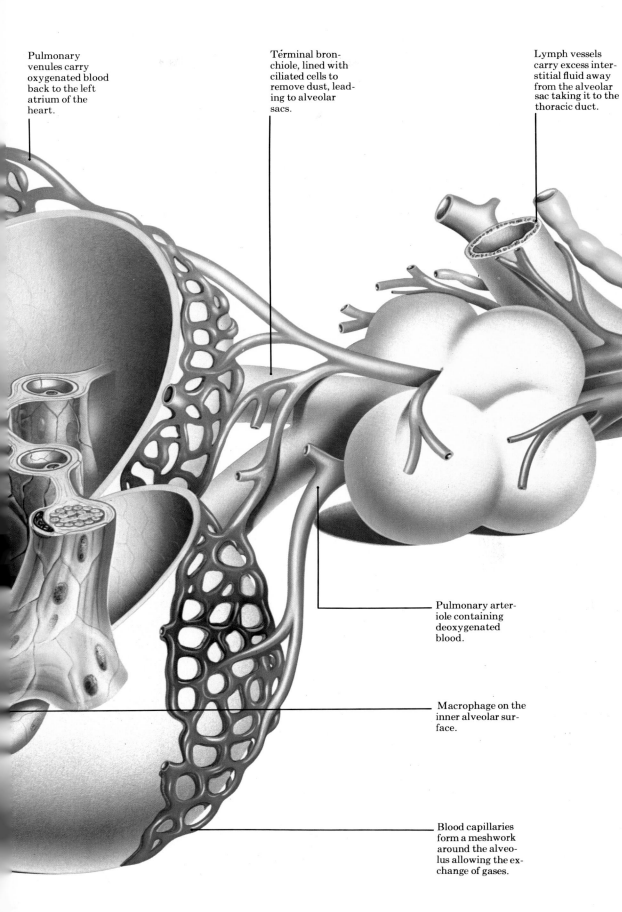

Pulmonary venules carry oxygenated blood back to the left atrium of the heart.

Terminal bronchiole, lined with ciliated cells to remove dust, leading to alveolar sacs.

Lymph vessels carry excess interstitial fluid away from the alveolar sac taking it to the thoracic duct.

Pulmonary arteriole containing deoxygenated blood.

Macrophage on the inner alveolar surface.

Blood capillaries form a meshwork around the alveolus allowing the exchange of gases.

Oxyhemoglobin Condition of hemoglobin when oxygen is combined with it.

Pharynx Cavity between the mouth and the posterior nasal cavity, above, and the beginning of the esophagus.

Phrenic nerve One of a pair of nerves, arising in the neck, supplying the diaphragm.

Pleura Sac-like structure that lines the inner surface of the thoracic cage and surrounds each lung.

Post-nasal cavity Dilatation of the back of the nasal passages, above the pharynx into which the Eustachian tubes open.

Pulmonary Refers to structures associated with the lungs.

Residual volume Air that remains in the lungs after forced expiration. This is 1,000ml.

Segment of lung Each lobe is subdivided into smaller sections, segments.

Sinuses Cavities within the skull that are lined with mucous membrane. They lighten the weight of the skull and act as resonant chambers for production of speech.

Speech centre Area of the brain concerned with the production of speech.

Tidal air Amount of air that is normally breathed in and out during quiet respiration, about 500ml.

Tonsil One of a pair of lymphoid structures found at the back of the mouth, at the entrance to the pharynx.

Trachea Windpipe from the larynx to bifurcation of bronchi, about 12cm long and kept open by C-shaped cartilages.

Vital capacity Total amount of air that can be breathed in and out of the lungs, 4,000ml.

Respiration 3

Asthma Difficulty in breathing due to contraction of the smooth muscle in the bronchioles. This gives a wheezing sound on respiration. There is a variety of causes, including allergy and infection.

Atelectasis Collapse of part of the lung, usually due to blockage of a small bronchus.

Bronchial carcinoma This is usually due to a cancer of the mucous membrane of a bronchus. It spreads to involve the local area of the lung and the local lymph glands. Cancer can spread from other parts of the body to involve the lung.

Bronchiectasis An area of the lung tissue and small bronchi that has been damaged by repeated infection, causing scarring and distortion of the tissues. This leads to problems with drainage of the normal secretions and a much increased likelihood of further infection.

Bronchiolitis Inflammation of the bronchioles.

Bronchitis This may be an acute infection of the bronchi that resolves without leaving any damage. Chronic infections may occur when the mucosal cells are damaged and the normal drainage cannot take place. This means that the infection is unable to clear up completely.

Cancer of the lung *See* Bronchial carcinoma.

Collapse of lung *See* Atelectasis.

Coryza Medical word for a cold.

Emphysema Condition of the alveoli in which their walls are damaged and broken down. This reduces the elasticity of the lungs and also the surface area over which exchange of the gases may take place.

Exchange of gases in alveolus

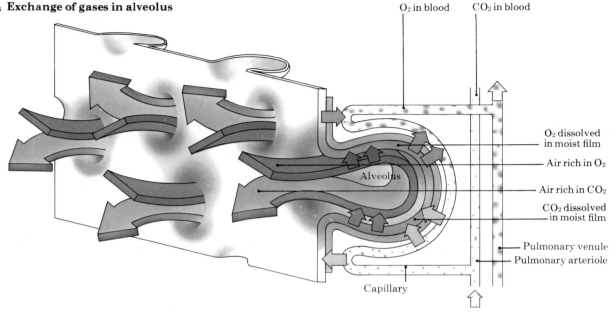

O₂ in blood — O_2 in blood
CO₂ in blood — CO_2 in blood
O₂ dissolved in moist film — O_2 dissolved in moist film
Air rich in O₂ — Air rich in O_2
Air rich in CO₂ — Air rich in CO_2
CO₂ dissolved in moist film — CO_2 dissolved in moist film
Pulmonary venule
Pulmonary arteriole
Alveolus
Capillary

The exchange of gases

The exchange of gases (*above*) has less than a second to take place. The carbon dioxide has to pass out of the plasma, where it is dissolved, through the capillary wall into the minute space between the capillary and the alveolar walls. It then passes through the alveolar walls into the thin film of moisture that lines each alveolus. Both carbon dioxide and oxygen have to dissolve in this moist layer on their way out of and into the blood. The gases move by diffusion —movement from an area of high to low pressure. The oxygen passes in the opposite direction to the carbon dioxide and combines with the hemoglobin in the red blood corpuscles to form oxyhemoglobin. The capillaries are so narrow that only one red blood cell at a time can get through. The oxygenated blood returns to the left atrium through the pulmonary venous system, which drains alongside the bronchioles and bronchi.

Inspired air contains 20 per cent oxygen, 0·03 per cent carbon dioxide and the rest is nitrogen. Expired air contains 16 per cent oxygen and the carbon dioxide is increased over a hundredfold to 4 per cent. In addition expired air is saturated with water vapour—this invisible loss of water from the body is about 1 litre a day.

Nervous control of respiration

The respiratory centre, in the medulla oblongata of the brain, automatically controls the breathing muscles through the phrenic nerve to the diaphragm and through the many intercostal nerves to the intercostal muscles. It can also make use of the accessory muscles of respiration when they are required.

The respiratory centre has two sources of information about breathing: The degree to which the lungs are stretched and the level of carbon dioxide in the blood.

The stretching of the lungs through inspiration is detected by branches of the vagus nerve which reflexly inhibit inspiration and allow expiration to take place. The fluctuations in the level of carbon dioxide are detected by nerve endings in the aorta and carotid arteries. An increase in carbon dioxide causes an increase in acidity of the blood and this stimulates the respiratory rate—hyperpnea.

Voluntary, rapid over-breathing decreases the carbon dioxide level and this is followed by a period of cessation of breathing—apnea. Permanently raised carbon dioxide levels, in some forms of lung and heart disease, lead the respiratory centre in the brain to adjust the respiratory rate to normal.

Breathing and swallowing

Normally the vocal cords are open as breathing takes place. When something is swallowed the soft palate is pulled up and closes the post-nasal cavity. At the same time the epiglottis closes the top of the glottis by muscles pulling up the larynx.

Involuntary breathing—cough

Irritation of the bronchi causes a deep inspiration followed by closing of the glottis. The expiratory muscles then contract causing increased pressure within the lungs. The glottis opens and an explosion of air, the cough, occurs.

Sneeze

Irritation of the nasal mucosa produces a reflex like that of the cough, but, as the air is violently expelled from the lungs, the tongue blocks the back of the mouth and the air passes through the nose.

Hiccup

A hiccup is due to a spasmodic contraction of the diaphragm causing the air to be rapidly inhaled. As this happens the glottis suddenly closes giving the characteristic noise.

Emotional breathing—laughing and crying

Both are long inhalations followed by short, sharp exhalations.

Yawning and sighing

Both are slow, long inhalations and gradual exhalation.

The organs of speech

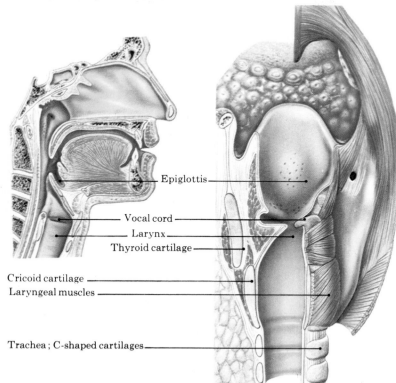

Epiglottis

Vocal cord
Larynx
Thyroid cartilage

Cricoid cartilage
Laryngeal muscles

Trachea; C-shaped cartilages

Speech and voice mechanism

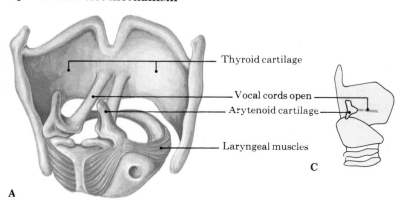

Thyroid cartilage

Vocal cords open
Arytenoid cartilage

Laryngeal muscles

C

A

Posterior views of larynx

Side views of larynx

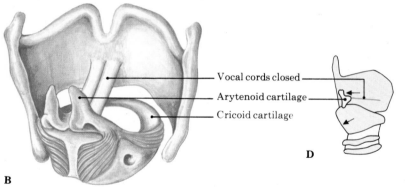

Vocal cords closed
Arytenoid cartilage
Cricoid cartilage

D

B

Speech and voice production

The larynx is the complex carti-lagenous structure between the pharynx and trachea. It divides the respiratory passage from that used for swallowing. It is commonly known as the "Adam's Apple" or voice-box. It has two functions—to prevent food and water entering the trachea by closing the glottis; and to produce sound.

The two vocal cords are mem-branes that run backwards inside the larynx (*left*). Normally they remain open and still during breathing (**A**). When they are drawn towards each other by the laryngeal muscles, the passage of air will make them vibrate and produce sounds (**B**). The faster the air passes them the louder the sound. The tighter the cords are drawn together the higher the note (**D**), the slacker, the lower (**C**). The tension in the vocal cords is altered by the laryngeal muscles tilting the small, posterior laryngeal cartilages and arytenoid cartilage.

Voice production is a highly com-plex matter of co-ordinating the breathing muscles, the vocal cords and the lips and tongue. The vocal cords can produce a wide range of musical notes. The resonance of these notes depends on the shape of the chest, mouth and sinuses. The finer qualities of speech depend on the shaping of the tongue and lips to give the characteristics of sibillants, consonants and labials in speech.

The voices of children of both sexes have a very similar range of notes. At the age of puberty a boy's voice "breaks". The larynx enlarges, due to the effect of the male hormone —testosterone—and the vocal cords become longer. This means that the boy is unable to produce high notes but now has a lower bass range. While the larynx is changing shape the quality of the voice also changes.

Speech depends on the brain. The production and quality of sound depends on the larynx and the sur-rounding structures to give reson-ance. The laryngeal nerves connect the "speech centre", in the cortex with the larynx. The speech centre is under the control of the cerebral hemispheres.

Laryngitis Hoarse-ness of the voice due to inflammation of the vocal cords and the larynx.

Pharyngitis In-flammation of the pharynx—back of the throat.

Pleurisy Inflam-mation of the pleura.

Pneumoconiosis Damage to the lung tissue from inhalation of dust particles, e.g. coal or stone dust.

Pneumonia Infec-tion of the alveoli causing impairment of breathing. This may either have spread from the bronchi, broncho-pneumonia, or occur in the whole of the lobe, lobar pneumonia.

Pneumothorax Air in the pleural cavity that may occur after the rup-ture of a lung.

Rhinitis Inflam-mation of the nasal passages. This may be due to a coryza or to allergy, hay fever.

Sinusitis Inflam-mation of the mucous lining of the sinuses.

Tonsillitis Infec-tion of the tonsils.

Tracheitis Infec-tion of the trachea.

Tuberculosis An infection of the lung caused by the organ-ism of tuberculosis. This is a chronic ill-ness that used to be a common occurrence.

Digestion 1

Definitions

Ampulla of Vater Point at which the common bile and pancreatic ducts enter the duodenum.

Amylase Enzyme that converts starch into maltose.

Anal sphincter Circular muscle at the end of the digestive tract.

Appendix Small finger-like protuberance from the cecum.

Bile duct Conveys bile from the cystic and hepatic ducts to the bile duct and into the duodenum at the ampulla of Vater.

Bile pigments Dark-coloured substances formed by the breakdown of red blood cells.

Bile salts Complex salts excreted by the liver that help in the emulsification of intestinal fats.

The digestive tract—the mouth

Food has to be broken down into its basic molecules before it can be absorbed into the body.

The mouth (*right*) has four functions: It breaks up food by chewing with the teeth and tongue; it lubricates it with saliva, to make swallowing easy; it regulates temperature by either cooling or warming the food; it consciously initiates swallowing when the bolus is ready.

There are three pairs of salivary glands: Sub-mandibular; sub-lingual; and parotid. Between them they produce 1,500 to 2,000 millilitres of saliva a day, containing the starch-reducing enzyme ptyalin.

An adult has 32 teeth: 8 incisors; 4 canines; 8 premolars; 8 molars; and 4 wisdom.

The tongue is an immensely mobile mass of muscle which helps the teeth to tear hard food into pieces by forcing it against the bony palate.

The throat and the esophagus connect the mouth with the stomach.

The mouth

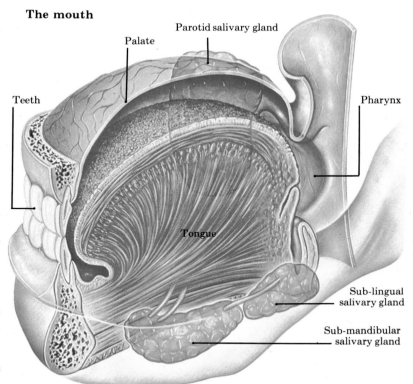

Palate
Parotid salivary gland
Teeth
Pharynx
Tongue
Sub-lingual salivary gland
Sub-mandibular salivary gland

The pancreas
Magnified about 4 times

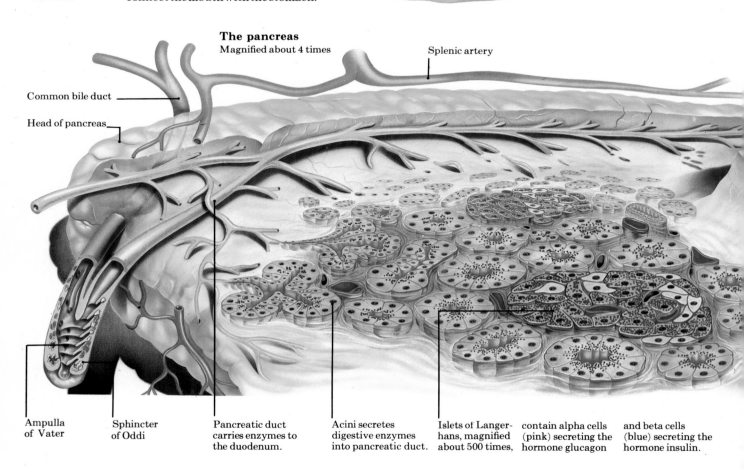

Splenic artery
Common bile duct
Head of pancreas

Ampulla of Vater

Sphincter of Oddi

Pancreatic duct carries enzymes to the duodenum.

Acini secretes digestive enzymes into pancreatic duct.

Islets of Langerhans, magnified about 500 times,

contain alpha cells (pink) secreting the hormone glucagon

and beta cells (blue) secreting the hormone insulin.

The digestive tract

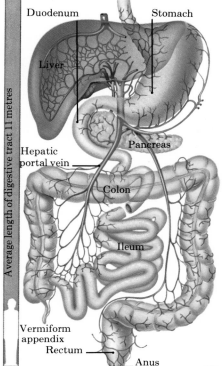

Duodenum — Stomach

Liver

Hepatic portal vein —

Pancreas

Colon

Ileum

Vermiform appendix

Rectum

Anus

Average length of digestive tract 11 metres

Tail of pancreas

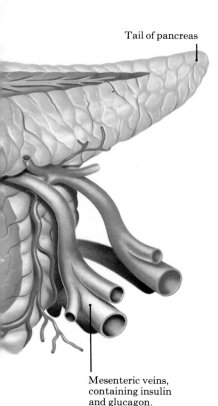

Mesenteric veins, containing insulin and glucagon.

The stomach

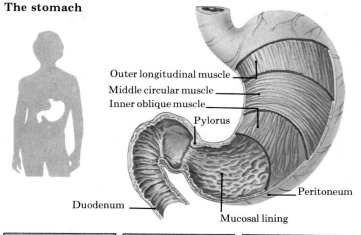

Outer longitudinal muscle

Middle circular muscle

Inner oblique muscle

Pylorus

Duodenum

Peritoneum

Mucosal lining

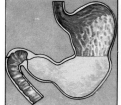

Peristalsis starts

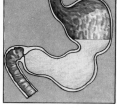

Pyloric sphincter opens

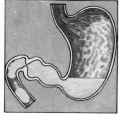

Chyme enters duodenum

The digestive tract

The digestive tract (*above left*) measures about 11 metres from mouth to anus. It is a strong muscular tube lined in the posterior part of the thorax with thick epithelium. It carries the bolus of food in a wave of peristalsis through the pyloric sphincter into the stomach.

The stomach

The stomach (*above right*) is a muscular sac that can contain about 1·5 litres of fluid. It lies in the upper part of the abdomen under the liver and left part of the diaphragm. It has three activities: The storage and gradual release of food into the duodenum; the physical activity that churns and squeezes the semi-liquid food, chyme; and digestion.

The stomach has six kinds of secretion: Mucus, to act as a protective layer; hydrochloric acid, to sterilize the contents, neutralize the salivary enzyme—ptyalin—and break down the inactive enzyme—pepsinogen—into pepsin; pepsin, which splits protein into peptones; rennin, which curdles milk; intrinsic factor, for absorption of vitamin B_{12} in the small intestine; and gastrin,

to maintain gastric secretions.

Although the stomach is essentially a storage organ it is able to absorb some water, alcohol and glucose. Small amounts of chyme are passed on at a time into the duodenum, where the main part of enzyme digestion takes place. These enzymes are formed by the intestinal wall and the pancreas.

The pancreas

In the pancreas (*left*) the head and body lie in the loop of the duodenum, and the tail stretches across to the spleen. It consists of a number of lobules with ducts merging into a main pancreatic duct that joins the common bile duct. It then enters the second part of the duodenum at the ampulla of Vater.

The pancreas has two kinds of secretion: Insulin, which is secreted into the circulation and controls carbohydrate metabolism; and enzymes that enter the duodenum.

There are three pancreatic enzymes: Amylase, to turn starch into maltose; lipase, to split fats into fatty acids and glycerol; trypsin, to split peptones and proteins into amino-acids.

Bilirubin Principal bile pigment.

Canaliculi Channels in the liver lobules into which bile is secreted.

Cholecystokinin (Pancreozymin) Hormone produced by the intestinal wall to stimulate the contraction of the gall bladder and secretion of pancreatic enzymes.

Cystic duct Tube from the gall bladder to join the hepatic and bile ducts.

Enterogastrone Hormone secreted by the duodenum to slow down gastric motility.

Enterokinase Secreted by the duodenum to change inactive pancreatic trypsinogen into trypsin.

Enterokinin Duodenal hormone that maintains the duodenal secretions.

Erepsin Duodenal enzyme to break peptones into amino-acids.

Fibrinogen Blood protein concerned with clotting.

Gastrin Hormone produced by the stomach to maintain the gastric secretions.

Glucose Simple sugar.

Glycogen Simple form of starch that can be broken down into glucose.

Hepatic duct Tube that drains the bile from the liver to join the cystic and bile ducts.

Ileocecal valve Valve that allows the chyme to pass from the terminal ileum into the cecum but not to return.

Insulin Glucose-controlling hormone secreted by the pancreas.

Intrinsic factor Substance secreted by the stomach that combines with vitamin B_{12}, extrinsic factor, so that it can be absorbed by the villi in the small intestine.

Digestion 2

The digestive hormones and autonomic nervous system

The digestive secretions depend on a complex relationship between the autonomic nervous system and the local production of hormones. The sight or smell of good food will produce reflex peristalsis, salivation and gastric secretion that are then maintained when it is eaten.

Similar hormones to gastrin, from the gastric mucosa, are produced in the duodenum. Secretin stimulates the pancreas to produce an alkaline secretion to neutralize the acid chyme from the stomach. Fats cause secretion of cholecystokinin—pancreozymin, which makes the gall bladder contract and the pancreas secrete enzymes. Enterogastrone slows down gastric motility and speed of emptying. The presence of food also produces enterokinin to maintain the duodenal secretions.

The small intestine

The small intestine (*right*) stretches from the pyloric sphincter to the ileocecal valve. It is about 7 metres long and is lined with small finger-like protuberances — villi — which are covered with even smaller ones—microvilli. This gives a total surface area of about 350 square metres for absorption. This area sheds about 125 grammes of dead cells a day. Most of this loss of protein is reabsorbed into the body, but the debris is excreted. These villi are able to cope with 12 litres of intestinal contents, food, water and secretions, a day. The successful absorption of nutrients depends on their transport away from the cells of the villi into the bloodstream or lacteals.

Small intestine
Magnified about 5 times

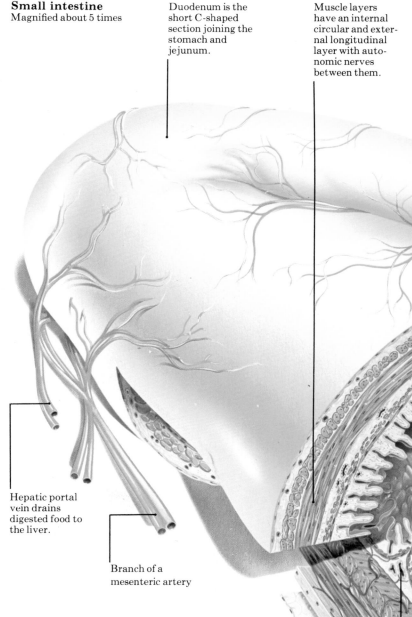

Duodenum is the short C-shaped section joining the stomach and jejunum.

Muscle layers have an internal circular and external longitudinal layer with autonomic nerves between them.

Hepatic portal vein drains digested food to the liver.

Branch of a mesenteric artery

Enzyme-producing glands

The villus

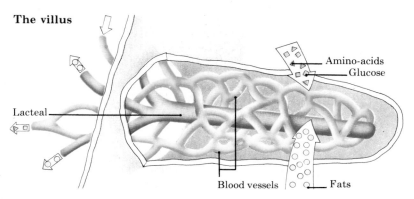

Lacteal

Amino-acids

Glucose

Blood vessels

Fats

The villi

Each villus (*left*) contains a network of blood vessels and a lymphatic vessel, the lacteal. The venous blood in the portal system takes the amino-acids, glucose, salts and water-soluble vitamins to the liver. The fatty acids and glycerol are combined together in the cells of the villus to form fats. These fats and the fat-soluble vitamins A and D drain into the lymphatic system via the lacteals in each villus.

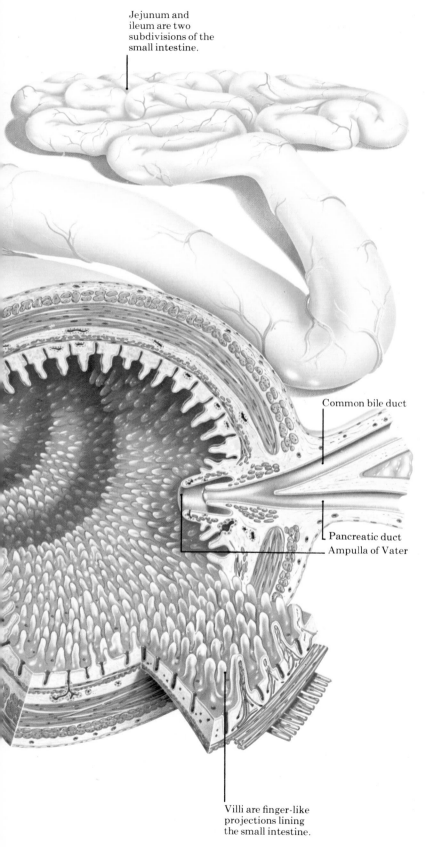

Jejunum and ileum are two subdivisions of the small intestine.

Common bile duct

Pancreatic duct
Ampulla of Vater

Villi are finger-like projections lining the small intestine.

The duodenum

The duodenum is the first C-shaped part of the small intestine and is responsible for the breakdown of food. The jejunum and ileum, extending from the duodenum, deal with absorption of food.

The duodenum produces three enzymes: Erepsin breaks peptones into amino-acids; invertase and maltase, to reduce sugar to glucose; enterokinase, to convert the inactive trypsinogen, the pancreatic secretion, to the active trypsin.

The bile salts, from the liver, help emulsify fats so that lipase can work more efficiently.

Intestinal musculature

The large and small intestine have a mucosal lining, inner circular and an outer longitudinal muscle coat. In the large intestine the outer layer is present only as three bands—the teniae coli. Peristalsis is controlled by the autonomic nervous system. The mucosa of the small intestine has patches of lymphoid tissue—Peyer's patches—which help eradicate infection.

The large intestine

The large intestine is about 1·5 metres long. It starts at the cecum, in the right iliac fossa, ascends to the liver, then crosses the abdomen, as the transverse colon, to the spleen before descending to the pelvis. It then narrows to become the sigmoid colon and finally the rectum before ending at the anus.

The vermiform appendix is attached to the cecum. It has no known use. The liquid ileal contents pass through the ileocecal valve into the cecum. As these waste products pass down the colon water and salts are reabsorbed. The fecal matter is further decomposed by bacterial action to produce gas, some of the vitamin B complex and feces.

Normally the rectum remains empty, but the feeling of distension on filling with feces causes defecation. This is a reflex action in babies that comes under conscious control as the nervous system matures. Rectal filling occurs due to peristalsis and the more fibre in the diet the softer the feces and the easier it is to eventually defecate.

Ptyalin Salivary amylase to break starch into sugars.

Pyloric sphincter Circular muscle between the stomach and the duodenum.

Rennin Gastric enzyme that curdles milk.

Secretin Duodenal hormone that stimulates the secretions from the pancreas.

Sinusoid Channel taking the mixed blood from the hepatic artery and portal veins over the hepatic cells to the central vein of a liver lobule.

Sphincter of Oddi Circular muscle adjacent to the ampulla of Vater closing the joint pancreatic and biliary ducts.

Trypsin Pancreatic enzyme that splits peptones and proteins into amino-acids.

Trypsinogen Inactive pancreatic enzyme that is changed into trypsin.

Urea Nitrogen-containing substance formed from ammonia by the liver.

Villus Finger-like protuberance that increases the surface area of the small intestine.

Vitamin B_{12} An essential vitamin for the formation of red blood cells, the extrinsic factor, that combines with the intrinsic factor to be absorbed in the small intestine.

Common diseases

Appendicitis Inflammation of the vermiform appendix that frequently leads to perforation and peritonitis.

Cancer May occur anywhere in the digestive system. It is rare in the small intestine but common in the mouth and esophagus, stomach and large intestine.

Cholecystitis Inflammation of the gall bladder.

Digestion 3

The liver

Lying mainly in the right side of the upper abdomen under the diaphragm and ribs, the liver (*right*) weighs about 1·5 kilogrammes. It is surrounded by a strong fibrous capsule that on the lower surface is covered with peritoneum. It has two lobes, a larger right and a smaller left. The gall bladder lies under the right lobe.

The liver contains 50 to 100,000 lobules. Each lobule has a central vein that drains blood into the hepatic veins. The portal veins and hepatic arteries are situated around the edge of the lobule and their blood intermingles to pass into the sinusoids that bathe the hepatic cells.

In the lobules, bile canaliculi drain bile into the branches of the bile duct, and thus into the hepatic duct. The bile is stored in the gall bladder, which contracts when stimulated by cholecystokinin. It then passes down the cystic duct into the common bile duct before joining the pancreatic duct at the ampulla of Vater.

The liver cells process the digested food, change it into substances that the body will need, then store them until they are required. If these substances are needed they will pass out of the cell into the sinusoid and thus to the central vein. Fats are either stored in the liver or sent back into the circulation of the fat depots. More frequently they are combined with protein—lipoprotein—to be used as an energy source. Proteins—the albumen, globulin and fibrinogen for the plasma—are all made in the liver. Glucose is turned into glycogen and iron combined with protein before being stored in the liver. The vitamins A, B complex, B_{12}, D, E and K are all also stored, the reserves lasting many months.

Liver cells also reprocess the body substances, such as hemoglobin, extracting the iron and re-using the amino-acids. Some of the amino-acids are changed into carbo-hydrates, to give readily available energy. The highly toxic substance ammonia is produced, which is turned into urea to be excreted by the kidneys. The heat produced by all this metabolism helps to keep the body temperature constant.

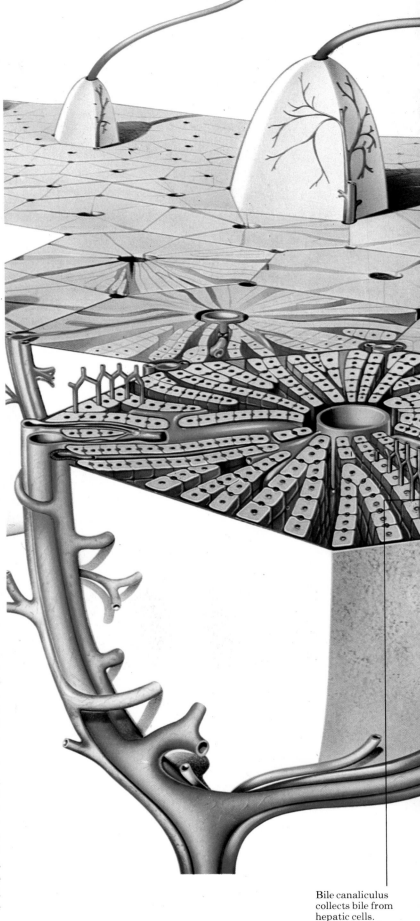

Bile canaliculus
collects bile from
hepatic cells.

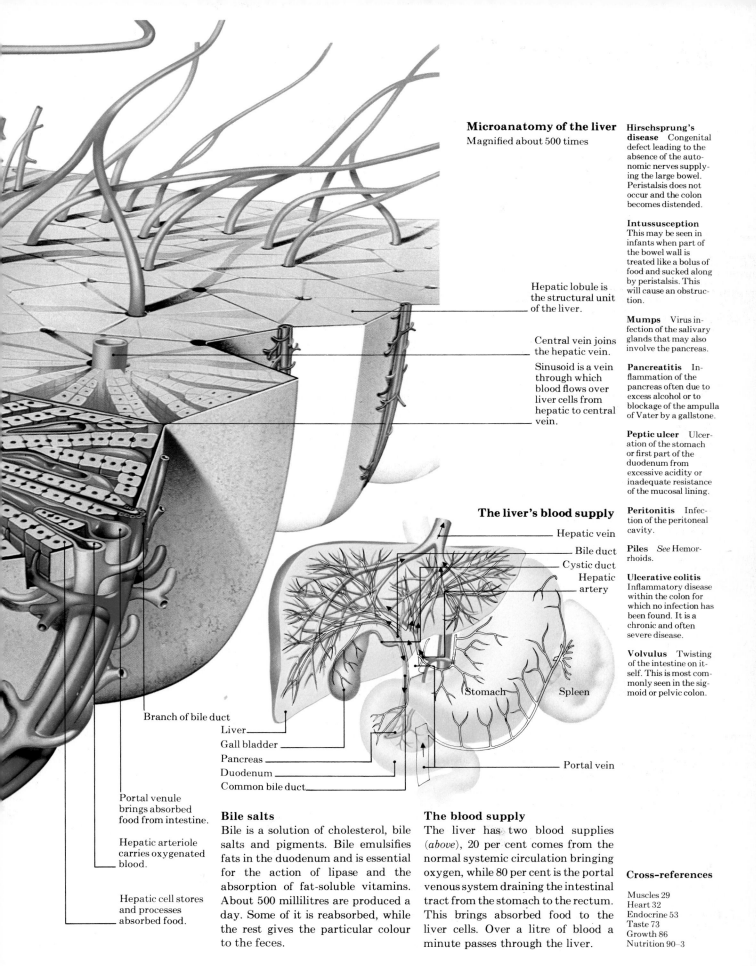

Microanatomy of the liver
Magnified about 500 times

Hepatic lobule is the structural unit of the liver.

Central vein joins the hepatic vein.

Sinusoid is a vein through which blood flows over liver cells from hepatic to central vein.

Branch of bile duct

Liver

Gall bladder

Pancreas

Duodenum

Common bile duct

Portal venule brings absorbed food from intestine.

Hepatic arteriole carries oxygenated blood.

Hepatic cell stores and processes absorbed food.

The liver's blood supply

Hepatic vein

Bile duct

Cystic duct

Hepatic artery

Stomach

Spleen

Portal vein

Bile salts
Bile is a solution of cholesterol, bile salts and pigments. Bile emulsifies fats in the duodenum and is essential for the action of lipase and the absorption of fat-soluble vitamins. About 500 millilitres are produced a day. Some of it is reabsorbed, while the rest gives the particular colour to the feces.

The blood supply
The liver has two blood supplies (*above*), 20 per cent comes from the normal systemic circulation bringing oxygen, while 80 per cent is the portal venous system draining the intestinal tract from the stomach to the rectum. This brings absorbed food to the liver cells. Over a litre of blood a minute passes through the liver.

Hirschsprung's disease Congenital defect leading to the absence of the autonomic nerves supplying the large bowel. Peristalsis does not occur and the colon becomes distended.

Intussusception This may be seen in infants when part of the bowel wall is treated like a bolus of food and sucked along by peristalsis. This will cause an obstruction.

Mumps Virus infection of the salivary glands that may also involve the pancreas.

Pancreatitis Inflammation of the pancreas often due to excess alcohol or to blockage of the ampulla of Vater by a gallstone.

Peptic ulcer Ulceration of the stomach or first part of the duodenum from excessive acidity or inadequate resistance of the mucosal lining.

Peritonitis Infection of the peritoneal cavity.

Piles *See* Hemorrhoids.

Ulcerative colitis Inflammatory disease within the colon for which no infection has been found. It is a chronic and often severe disease.

Volvulus Twisting of the intestine on itself. This is most commonly seen in the sigmoid or pelvic colon.

Cross-references
Muscles 29
Heart 32
Endocrine 53
Taste 73
Growth 86
Nutrition 90–3

Urinary System 1

The urinary system—the kidneys

The body has five organs of excretion: The skin for water, salts and urea; the lungs for carbon dioxide and water; the liver for bile salts, pigment and bilirubin; the intestine for roughage, water, salts and dead cells; and the kidneys. Each kidney is about 12 centimetres long, 3 centimetres thick and 7 centimetres wide, and weighs about 135 grammes, and is the most important route of excretion. A constantly adjusted trickle of urine is produced from the blood that enters each kidney, which is released when the bladder becomes full.

The kidneys filter 25 per cent of the blood supply at every heart beat. Blood under high pressure is forced into the capillaries of approximately 1 million glomeruli—Bowman's capsules—the goblet-shaped collecting cups of each filter unit—the nephron. The pressure forces 20 per cent of the plasma into the nephron, in the form of water, salts, urea, glucose and smaller molecules. Larger protein molecules and blood corpuscles cannot be forced through. The nephron then runs in a great U-shaped turn into the medulla. It consists of the proximal tube—loop of Henle—and distal tubule. It is surrounded by the blood vessels descending from the glomerulus. Two processes occur in the tubule: The lining cells reabsorb all the glucose, many of the salts and most of the water present in the filtrate; waste salts are secreted in the urine.

Urine is a pale yellow, slightly acid solution of salts, urea, uric acid, creatinine and metabolized hormones. Dilute urine will be less than the plasma's specific gravity of 1010, while concentrated urine may reach 1030. By selective excretion of acid or alkaline salts the slight alkalinity of the blood is kept constant.

The kidney has some part to play in the maintenance of blood pressure by the release of a hormone, renin, which increases blood pressure if the renal blood flow is reduced. It also regulates the body's water content by way of anti-diuretic hormone, which stimulates reabsorption of water from the filtrate into the blood.

The kidney Magnified about 6 times.

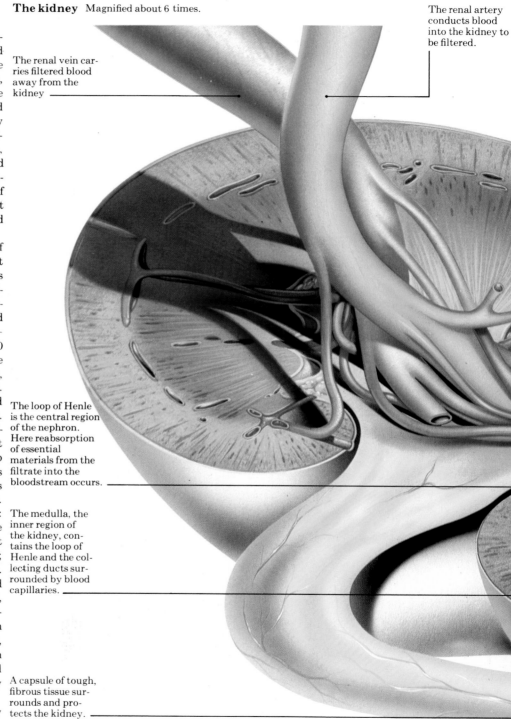

The renal vein carries filtered blood away from the kidney

The renal artery conducts blood into the kidney to be filtered.

The loop of Henle is the central region of the nephron. Here reabsorption of essential materials from the filtrate into the bloodstream occurs.

The medulla, the inner region of the kidney, contains the loop of Henle and the collecting ducts surrounded by blood capillaries.

A capsule of tough, fibrous tissue surrounds and protects the kidney.

The ureter is a muscular tube about 25 cm long, draining urine to the bladder.

The pelvis, at the core of the kidney, acts as a funnel, collecting urine from the nephrons and conducting it to the ureter.

An arteriole supplies the glomerulus in the kidney's cortex with blood under high pressure.

The glomerulus is a tight knot of blood capillaries from which water and chemicals filter into the nephrons.

The cortex is the outer region of each kidney. It contains the glomeruli of each nephron and the supplying blood vessels.

Bowman's capsule —the expanded beginning of the tubule—is tightly clasped around the glomerulus. It is the nephron's collecting cup for fluid filtering from the blood.

The nephron is one of over a million coiled tubules in each kidney. It extends through the medulla as the loop of Henle, ending in the collecting tubule. Each nephron collects filtered fluid from the Bowman's capsule.

Collecting duct from nephrons.

Capillaries surrounding loop of Henle.

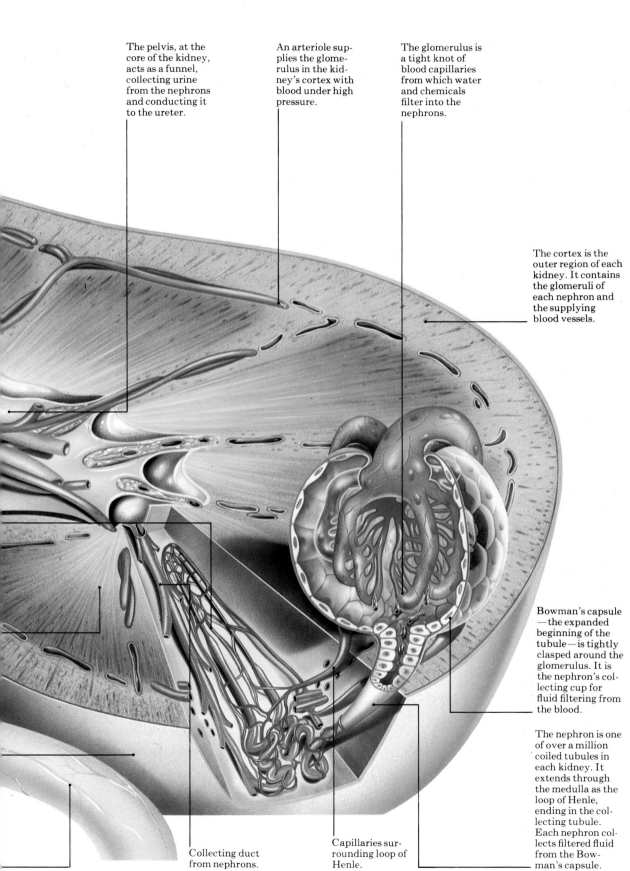

Urinary System 2

Organs of excretion

The urinary system (*below*) consists of two kidneys, the renal arteries and veins, two ureters and the bladder with the urethra.

The ureters, of circular and longitudinal peristaltic involuntary muscle, are 25 centimetres long, running behind the peritoneum and entering the posterior wall of the bladder at the base of the trigone.

The bladder, which can contain up to 500 millilitres of urine, lies in the pelvis, in front of the rectum and behind the symphysis pubis. In women it lies in front of the uterus. Partly covered with peritoneum, it is supported by the muscles and ligaments of the pelvic floor. The muscle is formed from strong, interwoven, longitudinal transitional and circular fibres and lined with epithelium; it is controlled by the sacral nerves and autonomic nervous system. The sphincter usually remains closed. The urethra is a fibro-elastic tube. In men it is about 20 centimetres long, passing through the prostate gland, pelvic floor and the corpus spongiosum of the penis. In women it is about 4 centimetres long, adjacent to the anterior wall of the vagina opening, behind the clitoris.

Organs of excretion

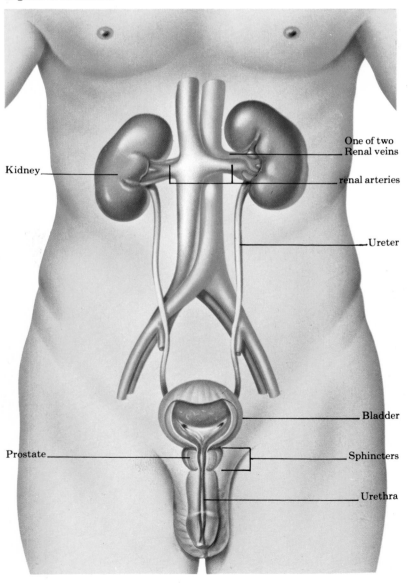

The filtering mechanism

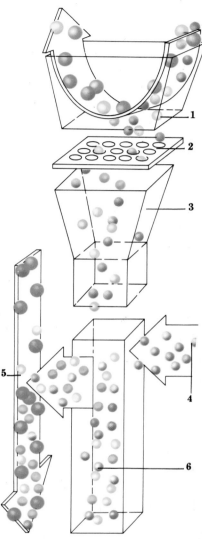

The filtering mechanism

Blood flows at high pressure (*above*) through the capillaries of the Bowman's capsule (1) and only small molecules are forced through the walls (2) into the first part of the nephron (3). The filtrate passes down the proximal tubule, which secretes further metabolites and salts (4) and reabsorbs water, sodium, essential salts, glucose and amino-acids into the blood (5). The loop of Henle and distal tubule are concerned with the reabsorption of water and maintenance of the overall acid-alkaline balance of the body. Unwanted salts, urea and water are left as urine (6).

Endocrine Glands 1

The endocrine system

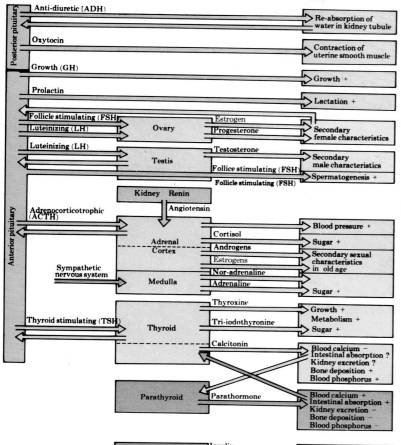

Stimulus
Feed back
+   Increased
−   Decreased
?   Effect uncertain

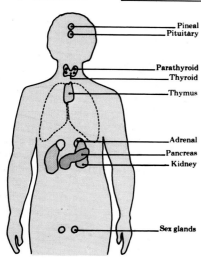

Pineal
Pituitary
Parathyroid
Thyroid
Thymus
Adrenal
Pancreas
Kidney
Sex glands

The endocrine system

Hormone responses, transmitted through the circulation, are much slower than those produced by the nerves. Each hormone stimulates or suppresses certain cells because it is able to fit into the cell's metabolism like a piece in a jig-saw puzzle. The response is detected by the hypothalamus in a "bio-feedback" system and this modifies the production activity of the "master gland"—the pituitary gland. As can be seen in the diagram (above) this gland controls many, but not all, of the remaining glands. Glands often interact through effects on the metabolism.

The lesser endocrine glands

The chart (left) shows the relationships between the various endocrine glands. Some of these have few or very local effects, such as the intestinal glands, while the activity of others, such as the pineal gland, is still not fully understood.

In the kidney renin is secreted in response to lowered blood pressure to combine with blood to form angiotensin, which stimulates aldosterone production.

In the intestine there is a variety of hormones with local effects: Cholecystokinin, enterogastrone, enterokinin, gastrin, pancreozymin and secretin.

The pineal body is a small structure adjacent to the roof of the third ventricle of the brain which may be associated with the development of testes and ovaries.

The thymus lies behind the upper part of the sternum and, until adolescence, plays an important part in the development of immunity by secreting a hormone to stimulate the production of lymphoid tissue and lymphocytes.

The pancreas produces exocrine digestive enzymes and two endocrine hormones, insulin and glucagon, from over one million islets of Langerhans. Insulin is produced by the beta cells in response to rising blood sugar levels and acts by increasing storage and use by the tissues while reducing liver production. It also reduces the use of fat by the body. Glucagon, from the alpha cells, responds to a low blood sugar by increasing glucose production. Glucose levels are also affected by the thyroid and growth hormones—adrenaline, nor-adrenaline and cortisol.

The testes respond to follicle stimulating hormone by the production of sperm and luteinizing hormone by increased testosterone levels.

In the ovaries follicle stimulating hormone causes maturation of the ovarian follicle with production of oestrogen. Luteinizing hormone produces the post-ovulatory secretion of progesterone. The balance is achieved by "bio-feedback"

Definitions

Adrenaline Hormone produced by the adrenal medulla.

Aldosterone Mineral-corticosteroid produced by the adrenal cortex.

Angiotensin Peptide produced by the production of renin and blood to stimulate aldosterone production and increase sodium reabsorption.

Antidiuretic hormone (ADH)—vasopressin Hormone produced by the posterior lobe of the pituitary gland that acts on the kidneys to increase the reabsorption of water.

Calcitonin Calcium-controlling hormone produced by the thyroid gland.

Cortisol Glucocorticosteroid hormone produced by the adrenal cortex.

Endocrine Gland that secretes a hormone directly into the blood stream.

Estrogen Female sex hormone produced by the ovary.

Exocrine Gland that secretes enzymes into a duct.

Glucagon Glucose-stimulating hormone produced by the alpha cells of the Islets of Langerhans.

Gluco-corticosteroid One of a group of adrenal hormones, principally cortisol, that stimulates carbohydrate metabolism.

Growth hormone Hormone produced by the anterior lobe of the pituitary gland.

Hydrocortisone See Cortisol.

Insulin Hormone produced by the beta cells of the Islets of Langerhans that reduces the glucose level in the blood.

Mineral-corticosteroid Adrenal hormones, mainly aldosterone, controlling salt reabsorption in the kidneys.

Endocrine Glands 2

The pituitary gland

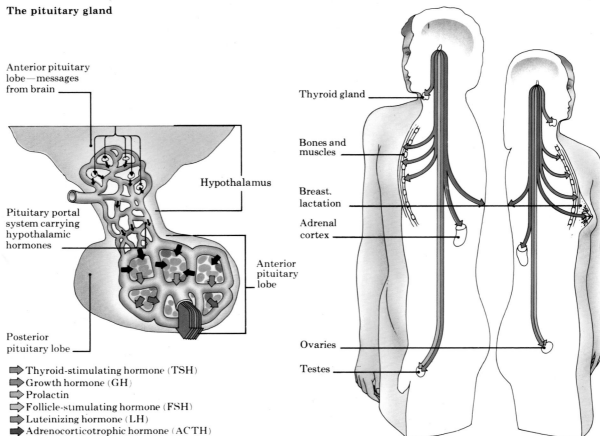

Anterior pituitary lobe—messages from brain

Hypothalamus

Pituitary portal system carrying hypothalamic hormones

Anterior pituitary lobe

Posterior pituitary lobe

Thyroid-stimulating hormone (TSH)
Growth hormone (GH)
Prolactin
Follicle-stimulating hormone (FSH)
Luteinizing hormone (LH)
Adrenocorticotrophic hormone (ACTH)

Thyroid gland

Bones and muscles

Breast. lactation

Adrenal cortex

Ovaries

Testes

The pituitary gland

The pituitary gland (*above*) lies under the centre of the brain, protected by the pituitary fossa—sella turcica—of the sphenoid bone. It is composed of two lobes, which measure 1 centimetre across and are connected to the base of the brain, just behind the optic chiasma, by a short stalk. The anterior lobe, about 70 per cent of the total gland, originally derives from the endoderm layer of the embryo and the posterior lobe from the ectoderm. This is reflected in their different functions.

The circadian biological rhythms —the emotional and physical changes—as well as the cerebral hemispheres, may stimulate or suppress the activity of the hypothalamus. The hypothalamus secretes controlling hormones into the pituitary portal venous system to increase or decrease hormonal activity in the anterior lobe.

The anterior lobe produces two kinds of hormones—those hormones that cause a direct response in the body and the trophic hormones that control the activities of the other endocrine glands. The level of stimulus from the trophic hormones is directly related to the response of the "target" gland and its "feedback" to the hypothalamus that assesses the response. This achieves a balanced hormonal activity.

Of those hormones that affect the body directly, growth hormone (GH) stimulates production of somatomedin by the liver which stimulates growth of bone, cartilage and possibly other body tissues. Prolactin helps to produce lactation and suppresses ovulation in the lactating woman. Thyroid stimulating hormone (TSH) increases thyroxine and tri-iodothyronine production from the thyroid gland. Follicular stimulating hormone (FSH) and luteinizing hor-

mone (LH) control the production of spermatozoa and testosterone in the testes, and estrogen and progesterone in the ovaries. Adreno-corticotrophic hormone (ACTH) stimulates gluco- and mineralo-corticosteroid production from the adrenal cortex.

The posterior lobe is like a physical extension of the hypothalamus as it is under nervous control. It secretes two hormones, which are stored until they are needed. The first, antidiuretic hormone (ADH), also known as vasopressin, acts on the tubule of the kidney nephron to increase the reabsorption of water. Any change in the osmotic pressure of the blood is assessed by the osmoreceptors in the hypothalamus; an increase causes stimulus to the thirst centre and release of ADH. There is an automatic increase in ADH secretion at night. The other hormone, oxytocin, stimulates the onset of labour, affects smooth muscle and helps in lactation.

Adrenal glands

The two adrenal glands (*right*) lie on the top of the kidneys and have a large blood supply as well as sympathetic nerves. There are two layers, the cortex and the medulla, through which a rich arterial supply reaches the underlying veins. Lying on the medulla of irregular cells surrounding the venous sinuses are long, straight columns of cells—zona fasciculata—followed by a layer of spherical cells—zona glomerulosa.

The adrenal cortex produces three kinds of hormones: Gluco-corticosteroids, mainly cortisol (hydrocortisone); mineralo-corticosteroids, mainly aldosterone; and a mixture of androgens and estrogens.

Cortisol helps stimulate glucose, amino-acid and fat metabolism to repair the body and increase levels of blood sugar.

Aldosterone monitors the level of body salts and causes reabsorption of sodium and loss of potassium from the kidney tubule. It also increases intestinal sodium absorption and decreases its loss in sweat. Angiotensin stimulates aldosterone.

Androgens and estrogens are produced in small amounts and continue

The adrenal gland

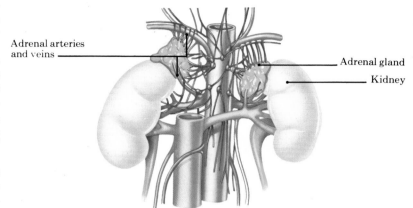

Adrenal arteries and veins

Adrenal gland

Kidney

to do so after the testes and ovaries cease developing. This feminizes men to a mild degree and masculinizes women when they are older.

ACTH stimulates the production of gluco-corticosteroids, and, to a far lesser extent mineralo-corticosteroids and sex hormones. It is controlled by the "bio-feedback"— the biological mechanism whereby the increase in production of a hormone causes a decrease in the production of the trophic hormone and thus reduces its own manufacture.

In stressful situations, such as emotional disturbance, pain, exposure to extremes of temperature or low levels of sugar in the blood, the adrenal medulla produces adrenaline and noradrenaline as a direct response to the sympathetic nervous system. The body is prepared for immediate action, "flight or fight", with increased blood pressure, shift of blood from the intestines to muscles, rapid heart rate, dilatation of bronchioles, sweating and increased alertness.

The parathyroid gland

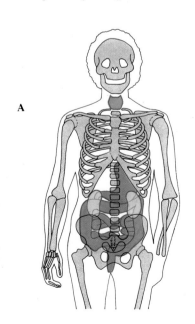

A

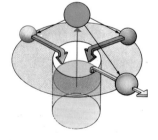

B

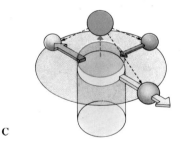

C

Thyroid gland

The two lobes of the thyroid gland (**A**), joined by the isthmus, lie adjacent to the upper part of the trachea. The gland produces thyroxine and tri-iodothyronine, that respond to TSH. They stimulate cell metabolism, increase blood sugar and are necessary for normal growth. It also secretes calcitonin which lowers blood calcium levels by stimulating deposition in the bones. It counteracts the action of parathormone and may affect absorption of calcium from the intestine and excretion in the urine.

Parathyroid glands

Four small glands are found in the substance of the thyroid gland (the black dots in **A**) and produce parathormone (PTH). This hormone regulates, with calcitonin, the levels of calcium and, indirectly, phosphate in the blood. Seen in **B**, low blood calcium (orange) increases the output of PTH to mobilize calcium from the bones (blue), inhibit secretion in the kidneys (red) and increase absorption from the intestines (green); this increases the blood calcium (**C**) and in turn inhibits the release of PTH while stimulating calcitonin.

Common diseases

Acromegaly and gigantism Due to an excess of growth hormone causing excess growth before maturity and thickening of bones and skull as an adult.

Addison's disease Deficiency of the adrenal cortical hormones leading to weakness, low blood pressure, vomiting, and eventual death.

Craniopharyngioma Cystic tumour of the pituitary gland.

Cretin Deficiency of thyroid hormones at birth leading to stunted growth and imbecility.

Cushing's disease Excess production of cortisol causing raised blood pressure, and a characteristic moonlike appearance to the face with obesity.

Diabetes insipidus Frequent passing of large volumes of urine due to deficiency of the antidiuretic hormone.

Goitre Swelling of the thyroid gland. This may be due to deficiency of iodine or excessive production of thyroid hormone.

Hyperthyroidism Excessive secretion of the thyroid hormones causing loss of weight, anxiety, rapid heartbeat and sweating.

Myxedema Deficiency of thyroid hormone leading to a slow mental state, thickening of the facial features, loss of hair and low body temperature.

Pheochromocytoma Tumour of the medulla of the adrenal gland with excessive production of adrenaline and noradrenaline.

Cross-references

Skeleton 22
Heart 33
Blood 37
Digestion 48
Urinary System 50
Nervous System 61
Sex 76–7
Pregnancy 80, 85

Nervous System 1

Definitions

Acetylcholine
Chemical produced in the transmission of nerve impulses.

Amygdaloid body
Part of the limbic system that regulates anger and aggression.

Arachnoid mater
Middle layer of the meninges.

Association areas
Parts of the cerebral hemispheres that relate one lobe with another.

Axon   Long fibre of a neuron.

Axon terminal
End of the axon.

Basal ganglia   Co-ordinate motor movements in association with the thalamus and cerebellum.

Brain stem   First part of the brain where the nerves cross before reaching the cerebral hemispheres.

Caudate nucleus
One of the basal ganglia.

Cerebellum   Part of the hindbrain that co-ordinates the movements.

Cerebral hemisphere   One of the two halves of the brain concerned with all higher mental functions.

Cerebrospinal fluid
Clear fluid bathing the brain and spinal cord.

Choroid plexus
Bunches of capillaries from which cerebrospinal fluid is formed in the lateral and fourth ventricles.

Cingulate gyrus
Part of the limbic system.

Corpus callosum
Connecting fibres between cerebral hemispheres.

Corpus striatum
Area next to the thalamus, in which basal ganglia are situated.

Dendrite   One of several small branches from the neuron that collects impulses from connecting neurons.

Anatomy of the nervous system

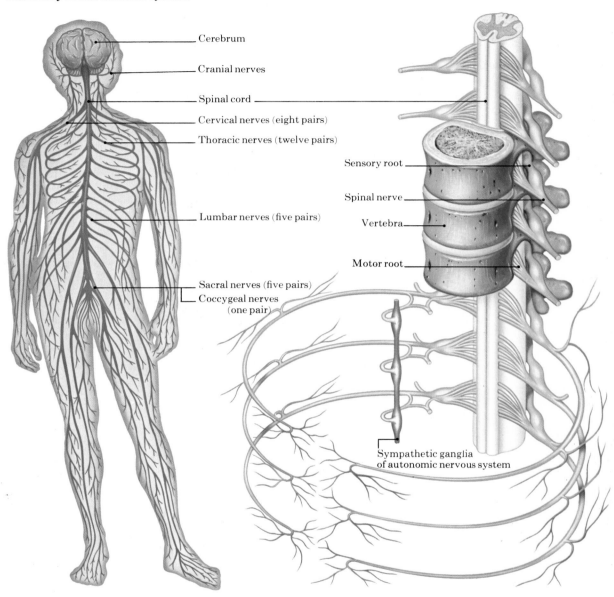

Cerebrum
Cranial nerves
Spinal cord
Cervical nerves (eight pairs)
Thoracic nerves (twelve pairs)
Lumbar nerves (five pairs)
Sacral nerves (five pairs)
Coccygeal nerves (one pair)

Sensory root
Spinal nerve
Vertebra
Motor root

Sympathetic ganglia of autonomic nervous system

The anatomy of the nervous system

The nervous system keeps control of all the body's activities. It has two parts, the central and peripheral nervous systems (*above*). The central nervous system consists of the brain and the spinal cord. The peripheral carries information to the central nervous system through the afferent, sensory nerves and carries out instructions through the efferent, motor nerves. The autonomic nervous system works outside the main nervous systems.

The brain—basic facts

The brain is a soft, jelly-like structure, weighing about 1,380 grammes in men and 1,250 grammes in women, surrounded by the meninges. It consists of ten thousand million neurons held in place by eighty thousand million neuroglia or glia.

There are three parts to the brain: The forebrain—the pair of cerebral hemispheres, thalamus, hypothalamus and limbic system; the midbrain—the first part of the brain stem; and the hindbrain—cerebellum, pons and medulla.

The meninges

The brain is enclosed by three membranes—the meninges—within the skull. The tough, outer dura mater and the thin, inner pia mater sandwich, the arachnoid mater. The arachnoid mater is pierced by many channels and allows the circulation of the cerebro-spinal fluid around the surface and to the ventricles and central canal of the spinal cord. The meninges, which also carry blood vessels, surround the nerves as they leave the brain and extend over the spinal cord down the neural canal.

Cranial nerves

There are twelve nerves, I–XII (*right*), which arise from the under-surface of the brain. They supply the head, neck and major organs in the body. Three are solely sensory (blue): Olfactory I; optic II; and auditory VIII. Two are entirely motor (red): Accessory to neck muscles XI; and hypoglossal XII, to the tongue. The rest are mixed: Oculomotor III; trochlear IV, and abducent VI, supply the eye muscles; trigeminal V, involves chewing and facial sensations; VII facial, involves facial expression and taste in the anterior two-thirds of the tongue; glossopharyngeal IX, involves swallowing and the remainder of taste; vagus X, supplies chest and abdominal organs.

Neurons

Neurons are the basic unit of the nervous system. Afferent neurons carry impulses from the sensory receptors, efferent neurons carry instructions to the muscles or glands, and interneurons link other neurons. A neuron's anatomy (*below*) consists of a cell body with nucleus and several small dendrites, which pick up messages. An axon connects with the next cell, where it breaks up into many small axon terminals. There is a gap—the synapse—between the axon terminal and the next cell. Neurons combine to form nerves.

Cranial nerves

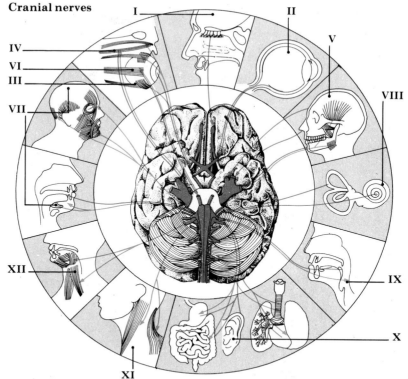

Nerve conduction

Normally the nerve cell is potentially excitable—polarized—with potassium ions inside and sodium ions outside the membrane. Once excited there is an exchange of ions—action potential—and a wave of electrical excitation will pass down the axon to the synapse to release noradrenaline or acetylcholine. The nerve is momentarily depolarized and while the ions are returning across the membrane it is in a "refractory" state. The synapse chemicals are rapidly destroyed by local enzymes; the synapse is then ready for action.

Dermatome Area of skin supplied by one nerve.

Dorsal root ganglion Afferent nerves containing neuron nuclei.

Dura mater Outer layer of the meninges.

Foramen of Magendie Hole in the roof of the fourth ventricle connecting the subarachnoid space with the ventricular system.

Foramen magnum Hole in the base of the skull.

Fornix Part of the limbic system.

Glia Supporting cells of the nervous system.

Grey matter Areas of the central nervous system containing many neuron nuclei.

Gyrus Fold of the cerebral cortex.

Hippocampus Part of the limbic system.

Homunculus Psychiatric term to describe a small man created in the imagination.

Hypothalamus Concerned with nervous control of the pituitary gland and vital nerve centres.

Lentiform nucleus One of the basal ganglia.

Limbic system Pair of structures concerned with memory and the instinctive emotions.

Lobe Anatomical division of the brain.

Mammillary body Initial relay station of the limbic system.

Medulla oblongata Part of the brain stem area connecting with the cerebellum.

Meninges Three layers coating the nervous system.

Motor end plate Specialized synapse between the axon terminals and muscle fibrils.

Neurons

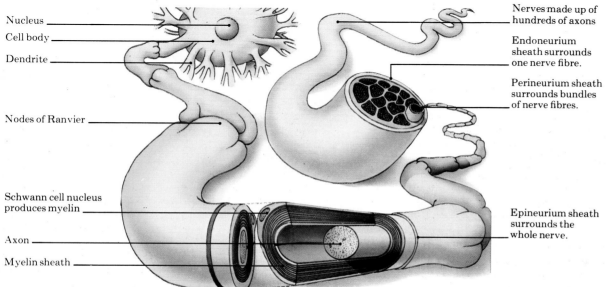

Nucleus

Cell body

Dendrite

Nodes of Ranvier

Schwann cell nucleus produces myelin

Axon

Myelin sheath

Nerves made up of hundreds of axons

Endoneurium sheath surrounds one nerve fibre.

Perineurium sheath surrounds bundles of nerve fibres.

Epineurium sheath surrounds the whole nerve.

Nervous System 2

Myelin Fatty coating of nerve.

Neural canal Bony channel in which the spinal cord lies.

Neuroglia See Glia.

Neuron Basic nerve cell.

Node of Ranvier Minute constriction in the myelin sheath of a nerve.

Nucleus Collection of nerve cells with one function.

Osmoreceptor Receptor sensitive to osmotic change.

Pia mater Inner, and thinnest, layer of the meninges.

Pons Area of interconnecting nerve fibres below the cerebellum.

Reflex arc Nervous response involving central nervous system.

Reticular formation Series of short nerve fibres whose activity increases alertness.

Schwann cell nucleus Concerned with the formation of myelin for the neuron.

Septum pellucidum Part of the limbic system.

Spinal cord Extending from the foramen magnum to the second lumbar vertebra with 31 pairs of nerves.

Sulcus One of 200 grooves between gyri.

Synapse Connection between the axon terminal and the adjacent cell.

Thalamus Concerned with sensory nervous transmission.

Ventricle One of four brain cavities.

Vital centre Nerve nuclei concerned with the normal functioning of the body.

White matter Area of the brain containing interconnecting neurons between areas of grey matter.

Spinal cord

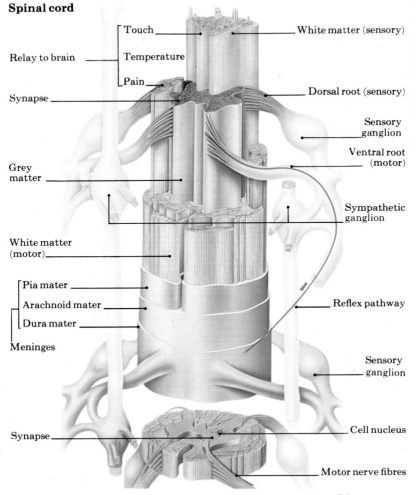

Touch — White matter (sensory)

Relay to brain — Temperature

Pain

Synapse — Dorsal root (sensory)

Sensory ganglion

Ventral root (motor)

Grey matter

Sympathetic ganglion

White matter (motor)

Pia mater

Arachnoid mater — Reflex pathway

Dura mater

Meninges

Sensory ganglion

Synapse

Cell nucleus

Motor nerve fibres

Skin sensation

Spinal nerves and reflex arc

The 31 pairs of spinal nerves (*right*) overlap each other in a way that developed in the embryo (*far right*). These nerves serve specific areas of skin known as dermatomes.

The reflex arc requires five things: Sensory receptor; afferent nerve; interconnecting neuron; efferent nerve; and end organ, such as muscle. Stimulation of the sensory nerve sends an impulse to the spinal cord; the impulse is relayed through at least one interconnecting neuron to activate the efferent nerve. Consciousness is not initially involved. This can be seen in the patellar reflex or emptying of a baby's full bladder. As the baby matures conscious inhibition can prevent the bladder reflex but not the patellar reflex.

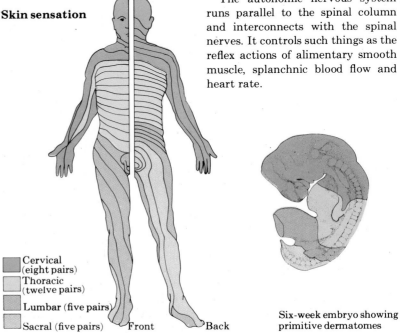

Cervical (eight pairs)
Thoracic (twelve pairs)
Lumbar (five pairs)
Sacral (five pairs)

Front   Back

Six-week embryo showing primitive dermatomes

The spinal cord

The spinal cord (*left*) extends from the foramen magnum to the second lumbar vertebra. It is covered with pia, arachnoid and dura mater and is bathed in cerebro-spinal fluid, which is also in the central canal. Sensory information reaches the cord through the dorsal nerve root—the dorsal ganglion. The posterior half of the central "grey matter" relays the afferent impulses to the columns of fibres in the surrounding "white matter". These columns have long fibres, ultimately reaching the brain, and shorter fibres connecting nearby segments of the cord. The lateral columns are concerned with pain and temperature and the posterior columns with touch and proprioception.

Some neurons transmit faster than others. The sensory nerves contain thicker myelinated fibres, for the fast transmission of pain at about 100 metres a second; non-myelinated fibres transmit heat or cold at slower speeds. The motor area is in the anterior half of the cord, with the central "grey matter" collecting instructions from the anterior columns and relaying them through the ventral nerve roots to the end organs. The dorsal and ventral roots combine to form 31 pairs of spinal nerves.

The autonomic nervous system runs parallel to the spinal column and interconnects with the spinal nerves. It controls such things as the reflex actions of alimentary smooth muscle, splanchnic blood flow and heart rate.

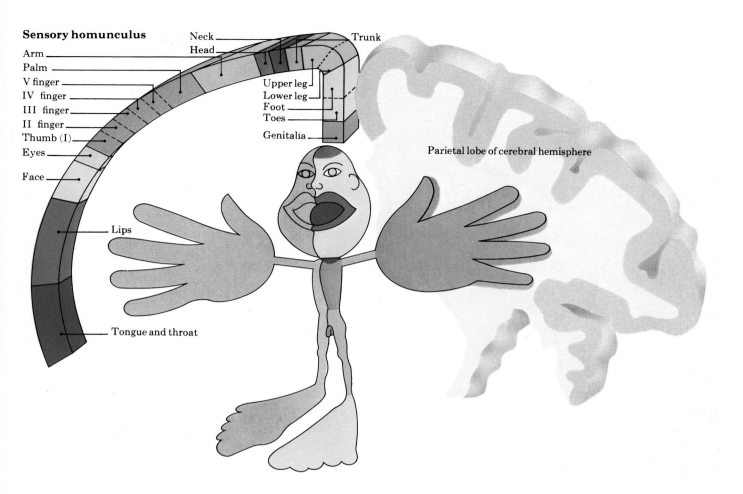

Sensory homunculus

Arm
Palm
V finger
IV finger
III finger
II finger
Thumb (I)
Eyes
Face

Neck
Head

Trunk

Upper leg
Lower leg
Foot
Toes

Genitalia

Parietal lobe of cerebral hemisphere

Lips

Tongue and throat

Cerebral cortex—homunculus

This weird picture of a man (*above*) is a visual representation of the relative proportions of the neurons in the sensory area behind the main sulcus of the brain. Skin areas on the trunk and limbs are poorly supplied with sensory nerves, while the lips, hands and feet are extremely sensitive and are relatively huge.

The visual cortex is in the occipital lobe and the auditory area in the temporal lobe so they are not proportionately represented. A similar but differently shaped homunculus can be made for the motor cortex. The hands are still huge but the face and tongue, trunk and legs are larger with much smaller feet.

This representation shows the right hemisphere, which controls the body's left side. The coloured areas on the diagrammatic representation correspond in size to the coloured areas on the body.

Intelligence

Intelligence is thought to be 80 per cent inherited and 20 per cent acquired. It can be approximately measured using the Stanford–Binet test in children up to the age of 15 and after that age the Wechsler test. Sex, race and cultural background have a variable bearing on the results. An average child who is well educated with a good understanding of words will appear more intelligent than is really the case. A rating of 100 is average and 130 very superior; below 70 is considered subnormal. There is little relationship between intelligence and creativity. Intelligence tests tend to assess the left cerebral hemisphere as this is the logical half of the brain, while creativity and originality are more from the right cerebral hemisphere. These are more difficult to measure as they depend on the subjective interpretation of the psychologist.

Learning

Learning is the way that new skills are acquired. Certain ways of doing something will produce a benefit or reward or, alternatively, something unpleasant so that repeating the action is avoided. It may be necessary to repeat the action several times before it is remembered.

Memory

Memory may be transient and immediately forgotten, such as not remembering everyone encountered in the street. Short-term memory is for things that are immediately necessary, such as shopping. Long-term memory can be "recalled" to consciousness when required.

Consciousness

Consciousness is constant activity of the cerebral cells by the excitation of the reticular formation. The subconscious is the activity of the memory part of the brain which may, unknowingly, modify the conscious.

Common diseases

Cerebral hemorrhage Bleeding from a cerebral artery into the surrounding brain tissue.

Cerebral thrombosis Blockage of a cerebral artery with blood clot.

Cerebral tumour Malignant or benign tumour may occur from glial cells or spread from other parts of the body.

Encephalitis Inflammation of the brain cells due to an infection.

Epilepsy An ill-defined episode of excessive activity in some part of the brain characterized by fits. Grand Mal—the major convulsive fit. Petit Mal—a momentary loss of awareness without loss of consciousness.

Nervous System 3

The brain

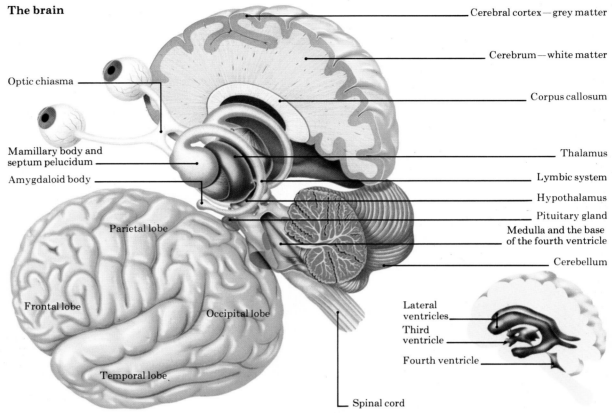

Optic chiasma

Mamillary body and septum pelucidum

Amygdaloid body

Parietal lobe

Frontal lobe

Temporal lobe

Occipital lobe

Cerebral cortex—grey matter

Cerebrum—white matter

Corpus callosum

Thalamus

Lymbic system

Hypothalamus

Pituitary gland

Medulla and the base of the fourth ventricle

Cerebellum

Lateral ventricles

Third ventricle

Fourth ventricle

Spinal cord

The forebrain—the cerebrum
Seventy per cent of the brain is formed by the two cerebral hemispheres (*above*). They are linked together by the nerve fibres running in the corpus callosum. The surface area is increased thirty times by the sulci and gyri. The cortex is densely packed with nerve cells, and the white matter with nerve fibres.

The nerve fibres cross over in the brain stem so that the right hemisphere is concerned with the left side of the body and vice versa. The left hemisphere is primarily the speech centre and mainly controls logical behaviour, but has to rely on the right side to assess a three-dimensional world, appreciate artistic values and recognize friends. The frontal lobe has conscious control of movement. The prefrontal area is concerned with intelligence and personality. The parietal lobe concerns sensation and body position. The smaller occipital lobe assesses vision and the temporal lobe assesses hearing. In between are the association areas linking the cerebral cortex.

The ventricles
Each cerebral hemisphere has a space—a lateral ventricle—looking like a "wish-bone". The two "arms" extend forwards into the anterior and temporal lobes and the "handle" into the occipital region. The lateral ventricles join together between the two thalami to form the third ventricle. A central duct leaves the fourth ventricle to pass down through the spinal cord. The fourth ventricle lies between the cerebellum and brain stem. The ventricular system is filled with cerebro-spinal fluid.

The cerebro-spinal fluid
Secreted by the choroid plexus, the cerebro-spinal fluid is clear and colourless. It bathes the surface of the brain in the subarachnoid space, between the arachnoid and pia mater, and internally in the ventricles, which it joins through a channel in the roof of the fourth ventricle—the foramen of Magendie.

It has three functions: To take food to the brain and spinal cord; to remove metabolites; to act as a shock absorber for the brain.

The basal ganglia or nuclei
The most important nuclei are the caudate and lentiform in the corpus stratum which, with the thalamus and cerebellum, produce smooth, precise movement. They help to co-ordinate motor movements.

The limbic system
Consisting of a pair of comma-shaped structures lying above the thalami, the limbic system links the midbrain with the hippocampus and cingulate gyrus and the cerebral cortex. It is concerned with instinctive emotions and memory. The mamillary body acts as the initial relay station through the fornix to the other areas. The septum pellucidum is concerned with pleasurable emotions, the amygdaloid body regulates anger and aggression. The hippocampus and cingulate gyrus help with memory by relating what is happening now with the memory of the things in the past. This helps to maintain the brain's concentration without being distracted by useless information. This inhibits the "alertness" created by the reticular formation.

The thalamus

Acting as a relay station for sensory information, the thalamus sorts out the impulses and sends them to the appropriate areas of the parietal lobes. It is also believed to be involved in assessing sensation.

The hypothalamus

Lying below the thalamus and above the pituitary gland in the centre of the brain, the hypothalamus is partly under the control of the prefrontal lobes and connected with the limbic system. It acts as a link between the nervous and endocrine systems, monitoring and regulating the autonomic systems as well as the metabolic state of the body.

It is connected with the anterior part of the pituitary gland by a portal venous system beginning and ending in capillaries; this is similar to the hepatic portal system. This blood carries the hormones released in the hypothalamus to control the release of the anterior pituitary hormones. The posterior pituitary gland is controlled by nerves.

The hypothalamus contains centres to control appetite, thirst, body temperature and libido. There is a fine balance between the amount of food eaten and used. Disturbance of the appetite centre may lead to excess appetite and weight gain.

If the blood becomes diluted the hypothalamus suppresses the production of the antidiuretic hormone (ADH) to allow diuresis to occur. Hemoconcentration will produce the opposite effect with stimulus to produce ADH from the posterior pituitary gland, and the sensation of thirst to stimulate the individual to drink. Libido is due to three factors: Instinctive responses, which are genetically inherited; the sex hormones; and sexual behaviour and responses learned from infancy. Hunger and thirst will reduce libido through its hypothalamic centre.

Temperature regulation is controlled through an assessment of the temperature of the blood. Heat loss can be increased by dilatation of the blood vessels in the skin, convection loss and sweating; it may be reduced by vaso-constriction and, if necessary, increasing heat production by shivering. The thyroid hormone, thyroxine, has a direct effect on the temperature control.

The activities of the hypothalamus integrate with the rest of the brain's activities, but are most closely correlated with those of the brain stem.

The hindbrain—the cerebellum

The hindbrain is joined directly to the spinal cord. It contains the medulla—concerned with sleep, consciousness, breathing and blood circulation—and the cerebellum.

The cerebellum grows rapidly and almost reaches adult size by the age of two. It co-ordinates the movements initiated by the rest of the brain. This is done by information relayed to it from the medulla oblongata via the nerves in the pons. An assessment is made of the conscious desire; information on balance is received from the ears, visual assessment from the eyes, and other sensory information from the rest of the body. A smooth, regulated movement can then be made, mainly by the inhibition of overreactions.

Areas and functions of the brain—memory, movement, emotion

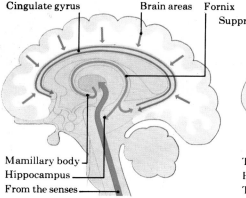

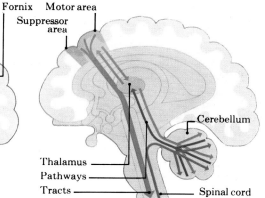

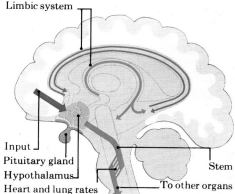

Memory

Memory (*above*) starts with sensation from the body passing to the cerebral cortex. As the information is relayed through the thalamus it is also passed to the mamillary body and into the limbic system. The stimulus passes through the fornix to the hippocampus and then outwards into the diffuse area of the cingulate gyrus. If the memory of a similar stimulus is aroused, the cerebral cortex may be activated.

Movement

Movement (*above*) is initiated in the motor cortex and immediately modified by the adjacent suppressor cortical area before direct transmission to the muscles. Each muscular movement is assessed and modified by the cerebellum in conjunction with the thalamus. An unconscious co-ordination of movement with position, balance and vision has to be made so that other muscles automatically adjust to give conscious movement.

Emotion

The emotions (*above*) are a complex integration of conscious reaction, memory and instinctive desires. The frontal lobes and limbic systems both affect the hypothalamus, with its centres for anger, thirst, appetite and sexual desire. These may be stimulated to interreact with the motor activity in the brain stem causing alterations in heart, respiratory rate and muscle tone. The hypothalamus also stimulates the pituitary gland.

Nervous System 4

The midbrain—the brain stem

The brain stem (*right*) is the vital control area of the brain and maintains all the essential regulatory mechanisms of the body: Respiration; blood pressure; pulse rate; alertness and sleep.

In the medulla oblongata—the lower end of the brain stem—the nerve fibres from the spinal cord, both sensory and motor, cross over.

At the upper end there are links with the higher cerebral centres of consciousness—the limbic and thalamic areas. The nuclei that reflexly control pupil size and eye movements lie in this area of the midbrain.

The third and fourth ventricles with the interconnecting central canal bathe the upper surface with cerebro-spinal fluid formed by the choroid plexuses. The central canal is joined to the subarachnoid space by the foramen of Magendie.

The pons connects the brain stem with the balancing control of the cerebellum. Short nerve fibres—the reticular formation—link the vital centres with each other and monitor the information reaching the brain stem from the sensory tracts of the spinal cord. A change in body position will cause an alteration in blood pressure and pulse rate by control of the diameter of the arterioles. The activity of the reticular formation maintains wakefulness and alertness. If this activity is slowed, for example after a large meal in quiet surroundings, sleep may occur. Anxiety or fear will increase alertness and raise the blood pressure and pulse rate.

The reticular formation not only controls the vital centres, as well as the production of hydrochloric acid in the stomach, but also helps correlate information from the eyes and ears, and smell and taste from the tongue. The sight and smell of a good meal will cause salivation, increased gastric secretion and peristalsis. The reticular formation then controls the mechanism of swallowing.

The vomiting centre may be activated by nauseating tastes or, in some people, sickening sights. It is easily stimulated by movement—motion sickness—due to the constant changes in the organs of balance.

Anatomy of the brain stem
Magnified about 3 times

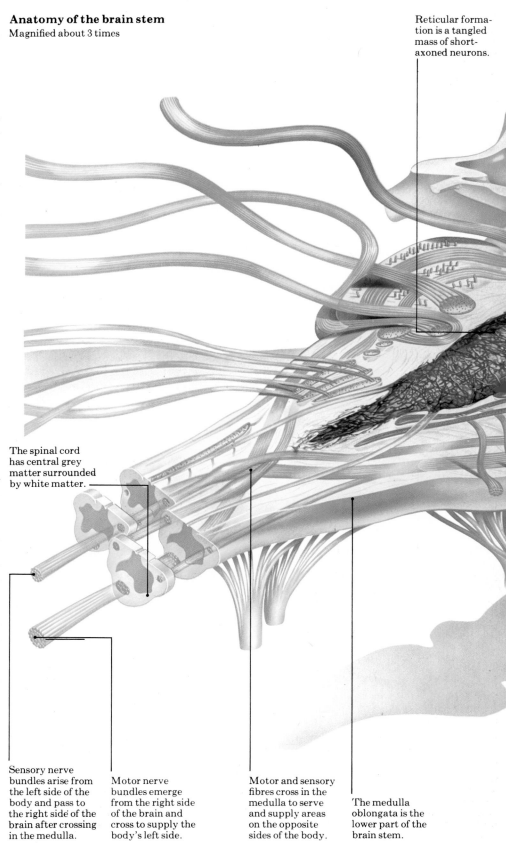

Reticular formation is a tangled mass of short-axoned neurons.

The spinal cord has central grey matter surrounded by white matter.

Sensory nerve bundles arise from the left side of the body and pass to the right side of the brain after crossing in the medulla.

Motor nerve bundles emerge from the right side of the brain and cross to supply the body's left side.

Motor and sensory fibres cross in the medulla to serve and supply areas on the opposite sides of the body.

The medulla oblongata is the lower part of the brain stem.

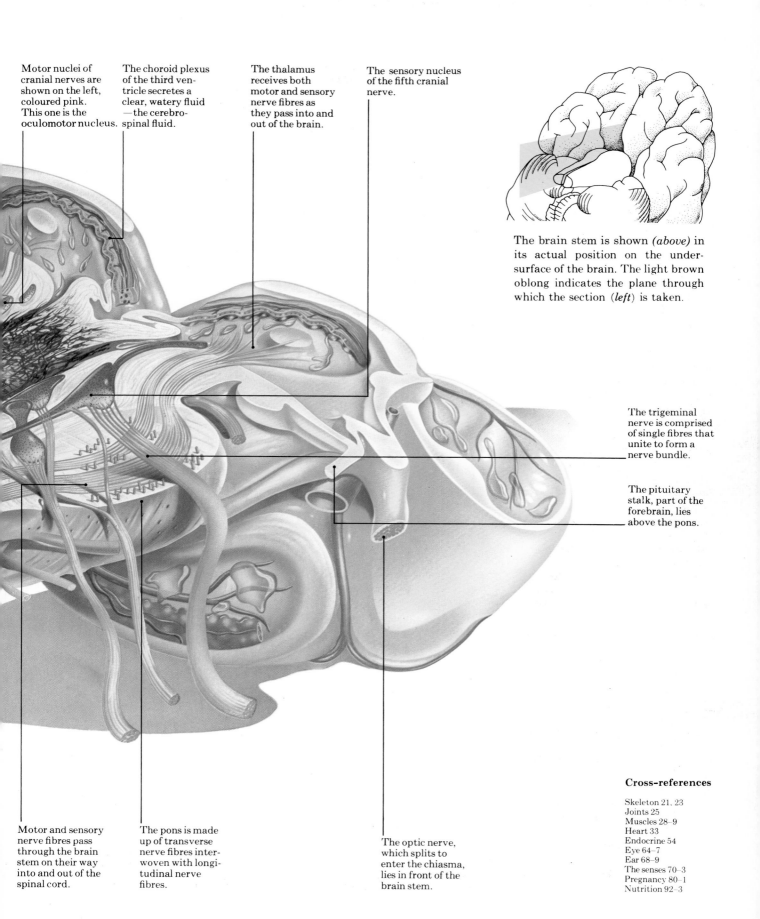

Motor nuclei of cranial nerves are shown on the left, coloured pink. This one is the oculomotor nucleus.

The choroid plexus of the third ventricle secretes a clear, watery fluid —the cerebro-spinal fluid.

The thalamus receives both motor and sensory nerve fibres as they pass into and out of the brain.

The sensory nucleus of the fifth cranial nerve.

The brain stem is shown (above) in its actual position on the under-surface of the brain. The light brown oblong indicates the plane through which the section (left) is taken.

The trigeminal nerve is comprised of single fibres that unite to form a nerve bundle.

The pituitary stalk, part of the forebrain, lies above the pons.

Motor and sensory nerve fibres pass through the brain stem on their way into and out of the spinal cord.

The pons is made up of transverse nerve fibres interwoven with longitudinal nerve fibres.

The optic nerve, which splits to enter the chiasma, lies in front of the brain stem.

Cross-references

The Eye 1

Definitions

Anterior chamber
Front part of the eyeball bounded by the cornea in front and the lens and iris behind.

Aqueous humor
Fluid contents of the anterior chamber.

Bipolar cell Cell with dendrites at both ends.

Blind spot Point at which the optic nerve enters the eyeball.

Canthus Angle where the eyelids join.

Cerebellum Part of the brain associated with co-ordination and balance.

Chiasma Area in which the two optic nerves cross.

Choroid Middle vascular coat of the eye.

Ciliary body Lies between the two chambers containing the ciliary muscle and ducts for draining the aqueous humor.

Ciliary muscle
Circular muscle in the ciliary body to which the suspensory ligament of the lens is attached.

Cones Retinal cells that respond to colour.

Cornea Transparent curved anterior part of the eyeball that helps to focus the light source.

Fovea Centre of the macula, packed with cone cells.

Geniculate body Relay station to the optic radiation.

Hyaloid canal Remnant of the fetal blood vessel that supplied the lens.

Iris Pigmented continuation of the choroid coat in front of the lens.

Lacrimal canaliculi Ducts that run from the inner canthus of the eyelids to the lacrimal sac.

Lacrimal gland Situated above the outer canthus.

Anatomy of the eye

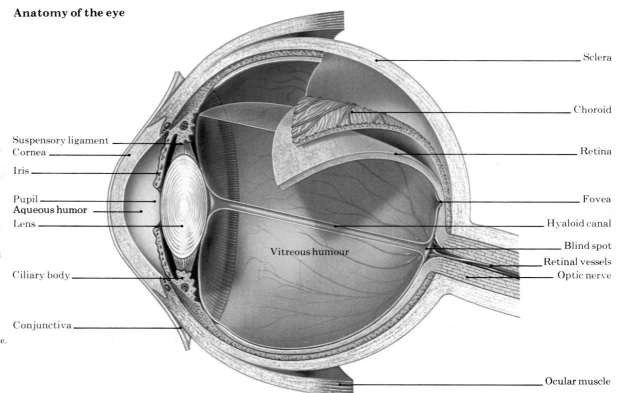

Suspensory ligament
Cornea
Iris
Pupil
Aqueous humor
Lens
Ciliary body
Conjunctiva

Sclera
Choroid
Retina
Fovea
Hyaloid canal
Blind spot
Retinal vessels
Optic nerve
Vitreous humour
Ocular muscle

The eyeball

The eyeball (*above*) is nearly spherical and more than 2 centimetres in diameter. It is a globular structure filled with a jelly-like fluid under slight pressure to give it firmness. The white outer surface—sclera—surrounds the eyeball except for the transparent cornea in the front.

The conjunctiva covers the outer surface of the cornea as it bulges forward. The aqueous humour is produced by the ciliary body and circulates through the posterior, behind the iris, and anterior chambers of the eye to bathe the inner surface of the cornea and lens.

The lens is a soft, bi-convex, transparent structure in a thin, tough capsule. It divides the anterior third of the eye from the posterior two-thirds, being held by the suspensory ligament to the ciliary muscle fibres.

The ciliary body contains the ciliary muscles, which alter the shape of the lens. It is close to the ducts that change the aqueous humour and, together with the iris and the choroid coat, forms the uveal tract. Much of the focusing of the eye is made by the

convex shape of the cornea; finer adjustments are made by the contraction of the ciliary muscles altering the lens shape.

The iris is the continuation of the choroid in front of the lens. Its colour, genetically determined, depends on the way in which the pigments are distributed. The pupil—the circular opening in front of the lens—varies in size very rapidly depending on the amount of light falling on the retina. The circular and radial muscle fibres in the iris are under the control of the autonomic nervous system. This prevents overstimulation of the retina by brilliant light. The pupil can vary in size from 1 to 8 millimetres. The vitreous humour fills the posterior two-thirds of the eyeball. Through the centre there is a thin, vessel structure that is the empty remnant of the foetal blood vessel that used to supply the lens in the foetus—the hyaloid canal.

The choroid coat lines the inner surface of the sclera and has brown pigmented cells to absorb light.

The optic nerve, containing 1 million nerve fibres, penetrates

the sclera and choroid coats and the nerves spread round the inner surface of the eyeball to form the retina. The point at which the optic nerve enters is known as the "blind spot" as there are no light-sensitive nerve cells at this point. The optic nerve is accompanied by an artery and vein that spread over the retina.

The retina consists of light-sensitive cells—the cones for red, green and blue, and the rods for shades from grey to white. The macula is the point at the back of the eye where the light will naturally fall when the eye is at rest. It has the greatest concentration of cones in the eye. The fovea is in the centre of the macula and is densely packed with colour-sensitive cones. It is responsible for the sharpest vision.

Visual acuity is the accuracy with which an object is seen; theoretically it must be of a size sufficient to stimulate one rod or cone. The two eyes work together—binocular vision—to relay the visual information to the visual cortex of the brain where the visual stimulus is interpreted in three dimensions.

Eye muscles—left eye

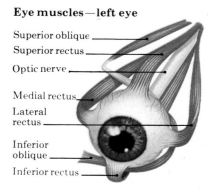

Superior oblique
Superior rectus
Optic nerve
Medial rectus
Lateral rectus
Inferior oblique
Inferior rectus

Eye movements

There are six muscles to control the movements of each eye (*above*); the movements are co-ordinated in the brain. If the lateral rectus muscle in one eye contracts, the medial rectus of the other will contract to a similar extent. The superior recti work together to pull the eyes back and to look up. The inferior recti make the eyes look down. The superior oblique muscle rotates the eye downwards and outwards and the inferior oblique upwards and outwards.

Pupil reflex

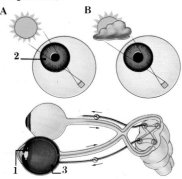

Pupil size and light intensity

The retina is very sensitive to light (*above*). Too much light (**A**) distorts colour and is dazzling. The pupils vary in size and thus reduce or increase the amount of light entering the eye. Bright light causes a reflex nervous reaction, controlled by centres in the midbrain. The circular pupillary muscle (1) in both irises contracts and the radial fibres (2) extend, thus narrowing the diameter. Poor light (**B**) will make both pupils dilate, allowing sufficient light to stimulate the cells in the retina (3).

The protective shield

The eyeball is deep set in the skull, under the orbital plate of the frontal bone and resting on the maxilla with the zygoma on the lateral side.

The hairy eyebrows, on the orbital ridge of the frontal bone, drain sweat and dirt to the side of the face. The long eyelashes of the upper lid prevent dust entering the eye by the automatic blinking reflex.

The upper eyelid contains a fibrous, tarsal plate that protects the closed eye when it firmly joins the softer lower lid. The upper lid is opened by the levator palpebra muscle and closed by the orbicularis oculi eye muscle, which also surrounds the lower lid. The edges of the lid are lined with modified sebaceous glands—the Meibomian glands—that keep the edges of the eyelid moist.

The conjunctiva is a thin, transparent membrane lining the inner surface of the eyelids and the outer surface of the eyeball; it acts as a protective layer and its smooth surface also allows easy movement of the eyelids.

The lacrimal apparatus

The lacrimal gland lies within the orbit just under the frontal bone as it joins the zygoma. This is at the lateral canthus. It secretes a salty fluid that lubricates the conjunctiva. This is drained by lacrimal ducts, from the inner corner of each eyelid at the medial canthus, into a dilated lacrimal sac. This in turn drains into the naso-lacrimal duct running down into the nose.

The lacrimal secretion has three main functions: To keep the conjunctiva moist; to act as a mild antiseptic; and to drain away dust and any small particles that may have landed on the conjunctiva.

Irritants, such as dust or infection, and emotion will cause a rapid increase in lacrimal secretion to such an extent that the drainage system is flooded and tears will overflow down the face. Blinking is a normal, reflex action occurring several times a minute to help lubricate the eye. It also occurs as a reflex, protective mechanism to sudden movement close to the face.

The retina

Light reaching the retina (*below*) has to pass through the nerve fibres—the ganglia (1)—joining them to the bipolar cells (2) that act as interconnecting neurons for the light-sensitive cells lying on the pigmented, retinal coat (3). The colour-sensitive cone cells (4, orange) are most densely congregated at the fovea (5). The rods (6, green) are found throughout the rest of the retina, giving wide black and white range.

The retina

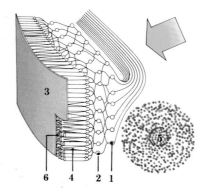

3
6 4 2 1
5

Rods

There are about 130 million rods and cones (photographed in detail below) in the ratio of 18 rods to 1 cone. The rods are more sensitive to light than the cones. There is a loss of colour vision and detail in poor light as there are fewer rods in the macula and none in the fovea. Light-sensitive, pigmented chemicals in the cones react to particular wavelengths of light—red, green or blue—giving a sensation of colour. The rods contain rhodopsin—visual purple.

Rods and cones

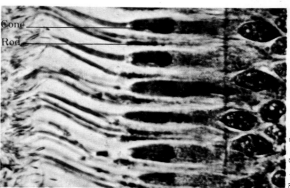

Cone
Rod

Photo: Gene Cox

Lacrimal sac Situated beside the inner canthus to collect fluid from the lacrimal canaliculi and drain through the naso-lacrimal duct.

Lens Soft, pliable, transparent, ovoid tissue that helps focus the light source.

Levator palpebrae muscle Raises the upper eyelid.

Macula Area of the retina where the light focuses on densely packed cells.

Meibomian gland Lubricating glands along the edges of the eyelids.

Naso-lacrimal duct Drains the lacrimal sac into the nose.

Neuron Nerve cell.

Oblique muscle Superior and inferior help rotate and move eye outwards and down or up.

Optic chiasma *See* Chiasma.

Optic nerve Cranial nerve (II).

Optic radiation Fibres from the geniculate body to the posterior cerebral cortex.

Optic tract Nerves from half of each eye from the chiasma to the geniculate body.

Orbicularis oculi muscle Surrounds both eyelids.

Posterior chamber Lies in front of the lens and behind the iris.

The Eye 2

Pupil Circular area in front of the lens surrounded by the coloured iris.

Rectus muscles There are four muscles in each eye—superior, inferior, lateral and medial.

Retina Layer of light-sensitive cells lying on the choroid coat of the eyeball.

Rhodopsin Light-sensitive pigment in the rods.

Rods Cells that respond to light but are unable to differentiate colour.

Sclera Tough white outer coat of the eye.

Suspensory ligament Ligament that suspends the lens from the ciliary muscle.

Tarsal plate Thin layer of fibro-cartilage in the upper eyelid.

Uveal tract Consists of ciliary body, iris and choroid coat.

Visual acuity The accuracy of vision.

Visual purple *See* Rhodopsin.

Vitamin A Combines with enzymes to help reconstruct the pigments that are broken down by light.

Vitreous humor Semi-fluid gel behind the lens.

Common diseases

Amblyopia Dimness of vision.

Astigmatism Disorder of focusing that disturbs the visual shapes of objects, e.g. a ball will look egg shaped.

Blepharitis Infection of the edges of the eyelids.

Cataract Opacity of the lens.

Chalazion Small round swelling of one of the lubricating meibomian glands in the edge of the eyelid which is caused by blockage of the duct in the glands.

The visual mechanism

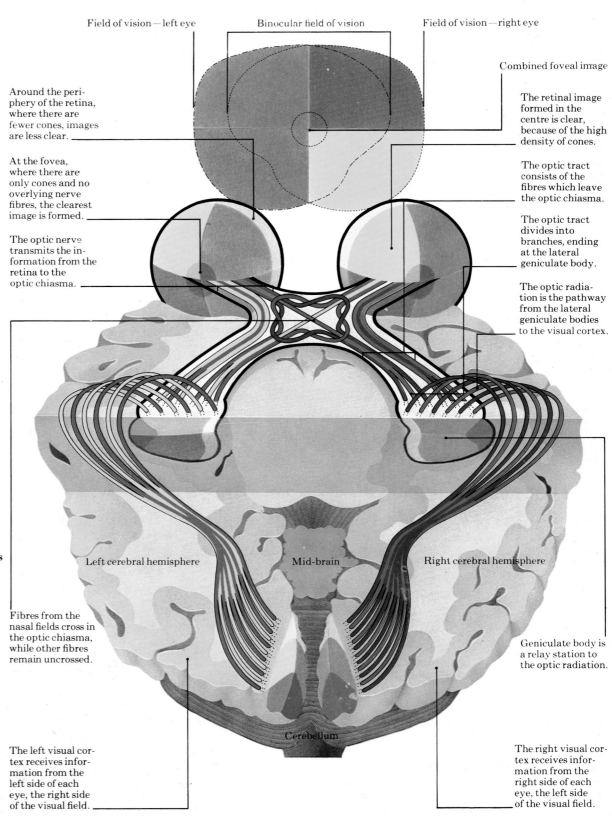

Field of vision—left eye

Binocular field of vision

Field of vision—right eye

Combined foveal image

Around the periphery of the retina, where there are fewer cones, images are less clear.

At the fovea, where there are only cones and no overlying nerve fibres, the clearest image is formed.

The optic nerve transmits the information from the retina to the optic chiasma.

The retinal image formed in the centre is clear, because of the high density of cones.

The optic tract consists of the fibres which leave the optic chiasma.

The optic tract divides into branches, ending at the lateral geniculate body.

The optic radiation is the pathway from the lateral geniculate bodies to the visual cortex.

Left cerebral hemisphere

Mid-brain

Right cerebral hemisphere

Fibres from the nasal fields cross in the optic chiasma, while other fibres remain uncrossed.

Geniculate body is a relay station to the optic radiation.

Cerebellum

The left visual cortex receives information from the left side of each eye, the right side of the visual field.

The right visual cortex receives information from the right side of each eye, the left side of the visual field.

Transmission to the brain

The cross-section of a brain (*left*), viewed from above, shows how the image that reaches the retina is coded and relayed to the visual cortex. Stimulation of the retina gives an image of various colours, shades and shapes to be transmitted in the optic nerve to reach the optic chiasma. At the chiasma half the neurons in each optic nerve change sides. This allows the brain to interpret the pictures from each eye in the correct way; the image on the outer half of one eye is the same as the nasal half of the other eye. The nerve fibres now continue in the optic tract to the geniculate body, which helps to co-ordinate the eye movements with the midbrain, cerebellum and the autonomic nervous system. The retinal image is then sent on through the optic radiation to the posterior lobe of the cerebral cortex, where, for the first time, the visual stimulus is interpreted in the consciousness in three dimensions. It is linked with other lobes by the association tracts.

Focusing mechanism

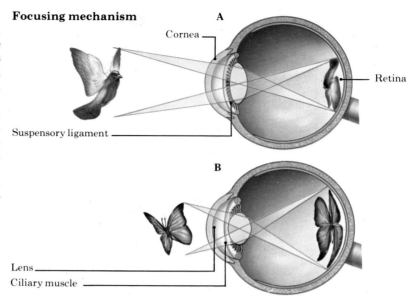

Cornea

Retina

Suspensory ligament

Lens

Ciliary muscle

Focusing mechanism

Focusing (*above*) is mainly dependent on the cornea and the relaxing and contracting effect of the ciliary muscles pulling on the suspensory ligament of the lens causing fine focusing. In distant vision (**A**) the muscles relax and the ligaments pull the lens into a disc shape. Close vision (**B**) requires a more circular lens, so the muscles constrict and the ligaments relax.

Colour blindness

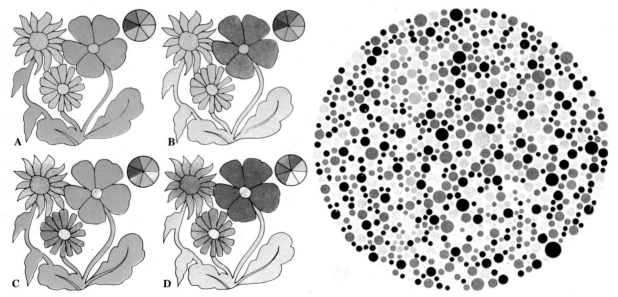

A   B

C   D

Colour blindness

Colour vision (*above left*) in its normal form (**A**) is thought to depend on the ability of three different types of cones to respond to red, green and blue light wavelengths.

About 10 per cent of all men and less than 1 per cent of all women have genetically inherited defective colour vision. This is commonly red or green (**B**) or, rarely, blue (**C**). Total colour blindness (**D**) can occur.

Defects in colour vision can be detected using Ishihara test plates, like the one above right. The spade is purple and thus stimulates blue and green receptors, and cannot be seen by the green colour blind. The fork is red so it is not seen by the red colour blind. Both are seen by those with normal colour vision.

Choroiditis Inflammation of the choroid coat.

Conjunctivitis Infection of the conjunctiva by virus or bacteria.

Diplopia Double vision.

Glaucoma Increased pressure within the eye that causes disturbance of vision and, if untreated, ultimate blindness.

Hypermetropia Long-sightedness.

Iritis Inflammation of the iris.

Meibomian cyst *See* Chalazion.

Myopia Short-sightedness.

Retina, detached Sheet of retinal nerve cells may, in part, lift off the choroid coat.

Retinitis Inflammation of the retina.

Retinopathy Degeneration of the retinal cells leading to amblyopia.

Squint Failure of both eyes to move in unison. This may be in all directions or, in the case of damage to one or more muscles, only in the direction in which those muscles operate.

Strabismus A squint.

Stye Infection of the hair follicle of the eyelid.

Uveitis Inflammation of the supporting mechanism of the lens and the iris.

The Ear

Definitions

Ampulla Dilation at the end of the semi-circular canal.

Auditory meatus Ear tube.

Ceruminous gland Modified sweat gland secreting cerumen.

Cochlea One of the two organs of hearing.

Corti, organ of Cells that lie in the scala media that detect vibrations in the endolymph produced by sound.

Endolymph Fluid in the scala media of the cochlea.

Eustachian tube Tube running from the posterior nasal space to the middle ear.

Foramen ovale Membrane at entrance of inner ear to which the stapes bone is attached.

Foramen rotunda Membrane covering the second hole in the wall of the inner ear.

Incus Middle of the three small bones of the ear.

Malleus Bone attached to the tympanic membrane.

Mastoid process Part of the temporal bone containing many air cells.

Otolith Minute crystals of calcium carbonate found on the hair cells of the utricle and saccule.

Perilymph Fluid in the semicircular canals, the scala vestibuli and tympani.

Pinna External ear.

Saccule Part of the organ of balance.

Scala media, tympani and vestibuli All parts of the cochlea.

Semicircular canals Three tubes, at right angles to each other, containing endolymph.

Stapes Smallest bone of the body attached to the foot of the foramen ovale.

The outer ear

In the ear (*right*) the outer ear consists of the pinna and external auditory meatus. The pinna is a thin sheet of cartilage moulded to direct sound towards the auditory meatus. It is covered with skin and, in some people, can be moved by weak muscles attached to the skull. The external auditory meatus is a blind tube 2·5 centimetres long ending at the tympanic membrane. In the outer third it is composed of cartilage and the skin is covered with protective hairs. The inner part lies in the temporal bone and contains modified sweat glands—ceruminous glands—that produce a waxy secretion—cerumen—to protect the skin and to drain out dust and bacteria.

The middle ear

The cavity of the middle ear lies deep in the temporal bone. The fibrous tympanic membrane stretches tightly across the end of the external auditory meatus. It is ovoid in shape, measuring about 9 millimetres across at its widest point, and transmits sound to the three small interlocking bones. The malleus is attached to the membrane and the incus links it with the stapes by two synovial joints. The stapes is attached to the foramen ovale of the inner ear. These bones vibrate in unison and magnify sound twenty times. The minute stapedius muscle, supplied by the VII cranial nerve, is attached to the stapes; the tensor tympani muscle, supplied by the V cranial nerve, tightens the tympanic membrane by its attachment to the malleus. Both muscles dampen excessive movement.

The air pressure in the middle ear is kept at atmospheric levels by the Eustachian tube, which connects with the posterior nasal space. The mastoid air cells, in the temporal bone, open into the posterior part of the middle ear.

The inner ear

The inner ear contains the cochlea, semicircular canals and auditory nerve.

The two ears, working in unison, can locate the source of a sound by the difference in volume and timing as it reaches the two sides of the head.

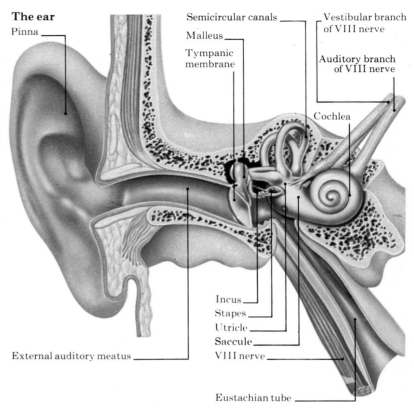

The ear
Pinna
Semicircular canals
Malleus
Tympanic membrane
Vestibular branch of VIII nerve
Auditory branch of VIII nerve
Cochlea
Incus
Stapes
Utricle
Saccule
VIII nerve
External auditory meatus
Eustachian tube

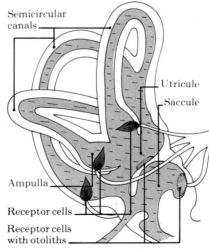

The organs of balance
Semicircular canals
Utricle
Saccule
Ampulla
Receptor cells
Receptor cells with otoliths

The organs of balance

The organs of balance lie deep in the petrous part of the temporal bone on either side of the skull adjacent to the cochlea in the inner ear (*above*). The three semicircular canals are at right angles to each other so any movement of the body will cause movement of the endolymph to stimulate the hair-like fibres, supplied with nerves, in the ampullae sited at the base of the canal. The endolymph will continue to move after rapid movement even when the body is stationary. The continued movement of the fine hairs by the fluid indicates to the brain that the head is moving but the eyes tell it that movement has stopped. This gives the sensation of vertigo.

Interconnecting with the ampullae of the three canals are the utricle and the saccule. They also contain hair cells with otoliths—minute crystals of calcium carbonate suspended in a jelly-like material. Pressure from the weight of these crystals stimulates the nerve endings when the head is stationary. Standing and lying will put pressure on different receptors. The vestibular branch of the VIII cranial nerve supplies the semicircular canals, utricle and saccule. Information is taken to the cerebellum, which co-ordinates movement and balance with the long sensory and motor tracts in the brain stem via the pons.

Organs of hearing

The two cochlea also lie in the petrous part of the temporal bones. Sound vibrations reach them through the foramen ovale from the stapes bones. Each is a narrow cone-shaped tube, 2·5 centimetres long, coiled two and a half times (*right*). They have three compartments: Scala vestibuli; tympani; and media.

Fluid vibrations of the perilymph pass up the scala vestibuli to the top of the cochlea and then down the tympani to reach the foramen rotunda in the medial wall of the middle ear. This allows the fluid to move by bulging when the pressure increases.

The human ear can detect sound between about 20 and 20,000 cycles per second (Hertz). The intensity of sound is measured in decibels (dB). A whispered voice is about 10 dB, a jet plane at take-off 110 dB. Decibels are a logarithmic measurement, so the difference between 40 and 50 dB is twice the volume, not 25 per cent as it might seem. Sound above 100 dB is damaging to the ear and it may lead to damage and deafness.

The cochlea

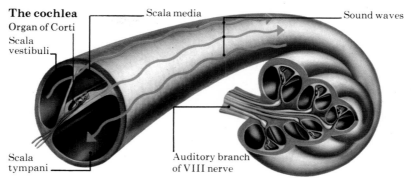

Scala media — Organ of Corti — Scala vestibuli — Scala tympani — Auditory branch of VIII nerve — Sound waves

The organ of Corti

In the scala media, which runs the full length of the cochlea and contains endolymph, lies the organ of Corti (*below*). It has five rows of cells on a basilar membrane, each with neat rows of hair-like projections. These cells are held in place by special supporting cells. The tips of the hairs are inserted into the gelatinous tectorial membrane, which overhangs them. Sound waves in the perilymph of the scala vestibuli and tympani cause the basilar membrane to vibrate at the appropriate wavelength. This stimulates the hair cells, which vibrate and activate one of the 30,000 neurons of the auditory part of the VIII cranial nerve. Sounds with more than one frequency will stimulate several areas of the organ of Corti. High frequencies activate the first part of the basilar membrane. The auditory nerve goes to the brain stem to reach the cochlear nucleus. Some fibres cross to the other side of the brain and interconnect with other nuclei; the rest pass to the auditory centre in the temporal lobe.

Tectorial membrane Membrane that overhangs the organ of Corti to detect movement of the endolymph.

Tympanic membrane Ear drum.

Utricle With the saccule forms part of the organ of balance.

Vestibular nerve Part of the VIII cranial nerve associated with balance.

Common diseases

Deafness Interference with sound reaching the cochlea owing to damage to the cochlea or sensory nerve or failure of conduction.

Ménière's disease Chronic condition due to intermittent raising of pressure of the endolymph of the ear.

Otitis externa Inflammation of external auditory meatus.

Otitis media Infection of middle ear.

The organ of Corti Magnified about 1,000 times

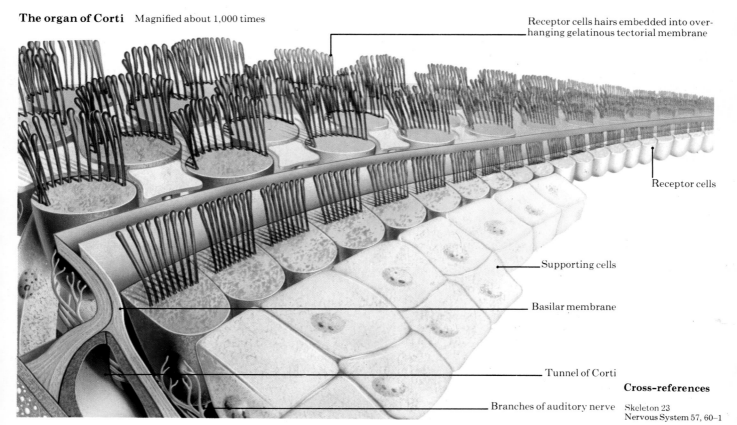

Receptor cells hairs embedded into overhanging gelatinous tectorial membrane

Receptor cells

Supporting cells

Basilar membrane

Tunnel of Corti

Branches of auditory nerve

Cross-references

Skin, Hair and Nails

The skin—the protective coat

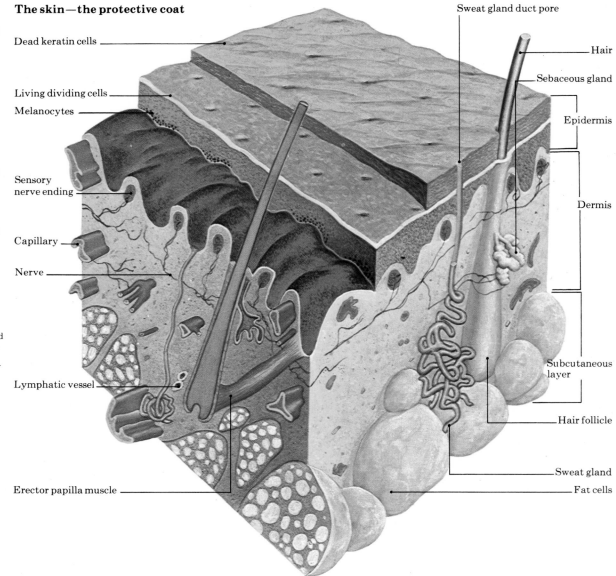

Dead keratin cells
Living dividing cells
Melanocytes
Sensory nerve ending
Capillary
Nerve
Lymphatic vessel
Erector papilla muscle
Sweat gland duct pore
Hair
Sebaceous gland
Epidermis
Dermis
Subcutaneous layer
Hair follicle
Sweat gland
Fat cells

The skin

The skin (in detail above) covers about 1·7 square metres and weighs about 3 kilogrammes. It is waterproof and a defence against trauma and infection; it is also a sensory organ and a temperature regulator. The outer layer of the epidermis is of dead, keratinized cells, which flake away all the time. Beneath this the epidermal living cells are growing outwards as they gradually form the keratin.

Underneath is the dermis, consisting of a network of collagen tissue fibres with interweaving blood and lymph vessels, sweat and sebaceous glands, and hair follicles and nerve endings; the latter are adapted to detect pain or pressure, heat or cold and vibration. Below the dermis is a variable layer of fat-storage cells.

The skin colour depends on the degree of melanin. All races have the same number of pigment cells—melanocytes—but genetic differences control the amount incorporated in the epidermal cells. Sunlight stimulates further production. Albinism is due to an absence of pigment-forming enzyme. About 3 million eccrine sweat glands in the skin produce water, waste salts and urea. In the ear they are modified into ceruminous glands to produce cerumen—wax. A different type of sweat gland—apocrine—occurs in the axillae and pubic areas. These secrete a thicker fluid that the skin bacteria breaks down to give body odour. Sebaceous glands, adjacent to the hair follicles, produce sebum—a fatty secretion that oils the hair and lubricates the skin by keeping the sweat on the epidermis.

The breast

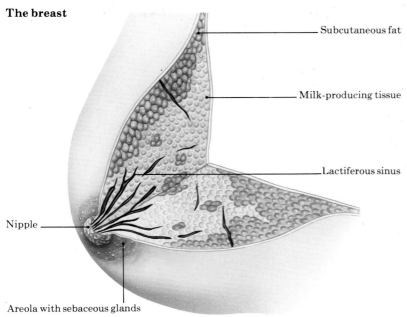

Subcutaneous fat

Milk-producing tissue

Lactiferous sinus

Nipple

Areola with sebaceous glands

The nail

Lunular pale crescent of active nail growth

Nail formed of keratin

Bone—one of the distal phalanges

Cuticle fold of epithelial cells

Nail bed formed from living layer of epidermis

Hair

The hair follicle in the dermis is a tube with a central core of dead, keratinized cells, which are pushed upwards as new cells are produced to form a hair. Each follicle has a small erector muscle that can contract to make the hair stand upright. Only the palms of the hands and soles of the feet have no hair. There are one to 200,000 hairs on the scalp which grow about 2 millimetres a week. The colour of the hair is dependent on its melanin content; the hair becomes whiter with age due to less pigmentation and minute air bubbles in the shaft. Balding, in part, is due to inheritance and, in part, to the male hormone—testosterone—produced in the testes.

The breast

The breast is composed of 15 to 20 modified sweat glands developing into lobes. They are rudimentary in the male. The female breast gland (*top left*) develops rapidly at puberty with secreting cells responding to the hormones in the menstrual cycle. During pregnancy the glands become congested and milk is collected in the lactiferous sinuses, which join behind the areola of the nipple. The areola is lubricated by the moist secretion of sebaceous glands.

The nails

The nails (*centre left*) develop at the third month of fetal life from a special layer of cells that fills with hard keratin and grows over the epidermis at the ends of the fingers and toes as a protective layer.

Temperature control

The temperature centre in the hypothalamus controls heat loss and production by the body through the skin (*below*). Overheating (**A**) causes an increased blood flow, from the blood vessels (1), to radiate heat and causes sweating, through the sweat glands (2), to lose heat. A fall in body temperature (**B**) constricts the surface blood vessels, stops sweating and makes the erector muscles (3) contract, causing the hairs (4) to stand on end, trapping air as an insulating layer. Additional heat can be produced by shivering.

Temperature control

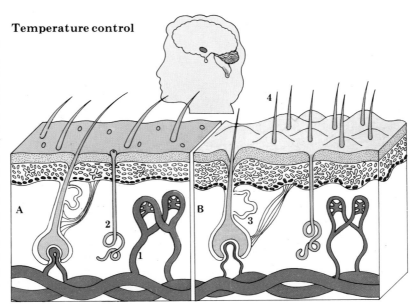

Carbuncle More severe form of boil with several points of discharge.

Dermatitis Inflammatory reaction of the skin that may be due to infection, contact with an allergy-producing substance or as a reaction to some internal disease.

Eczema Red, rough skin usually taken to refer to atopic eczema, an inherited condition in which there is roughness of the skin, particularly in the flexures of the arms and legs, but sometimes the whole skin.

Folliculitis Staphylococcal infection of the hair follicles.

Impetigo Rapidly spreading infection of the skin that may be caused by streptococci or staphylococci.

Keloid scar Overgrowth of the healing fibrous tissue of the scar that causes a hard, slightly raised, reddened surface.

Moles Raised, slightly pigmented areas of the skin found in most people that only very rarely become malignant.

Paronychia Infection at the nail bed.

Pediculosis Infestation of the scalp or pubis with lice.

Prickly heat Blockage of the sweat glands occurring after prolonged sweating in a hot climate.

Seborrheic dermatitis Scaling redness of the scalp, eyebrows, axillae and the face. It is due to an inborn tendency and may be aggravated by mild infection.

Tinea Fungal infection of skin.

Verruca Viral wart.

Cross-references

Muscles 27
Nervous System 60–1
Pregnancy 80, 85

Touch, Taste and Smell

Skin sensation

Some areas of the skin are densely packed with nerve endings, as in the finger-tips, while others, as on the back, have comparatively few. This is reflected in the proportion of the parietal cortex devoted to various areas of the body.

The nerve endings which are concerned with touch, pressure and slight variations in temperature have similar structures. Slight differences in both structure and position have led to the nomenclature illustrated (*right*). Nerve endings in the hair follicle detect hair movement.

Many nerve endings do not have end organs and these detect light, touch and pain. Sufficient stimulus will cause pain in any nerve ending. The brain correlates all skin sensations with the eyes, ears and nose.

Skin sensation

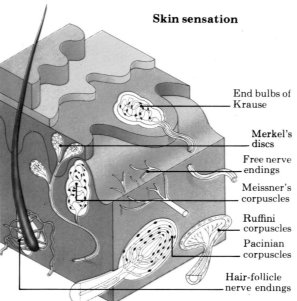

End bulbs of Krause

Merkel's discs

Free nerve endings

Meissner's corpuscles

Ruffini corpuscles

Pacinian corpuscles

Hair-follicle nerve endings

Referred pain

Internal organs and structures are well supplied with nerves, but pain is diffuse and poorly located compared with skin sensation (*below*). Most of the pain is produced by stretching and contracting, hence the pain of colic. Internal pain will cause stimulation of local nerves in a segment of the spinal cord; this makes it appear that the pain is coming from the skin, which is supplied by the sensory nerves. The heart (1) and oesophagus (2) refer pain to the neck, shoulders and arms; uterus (3) and pancreas (4) to the lumbar region; and kidneys (5) into the groin. Diaphragmatic pain may be referred to the shoulders, as the phrenic nerve of the diaphragm is formed from the spinal nerves in the neck, which also supply the shoulders.

Smell—the nose

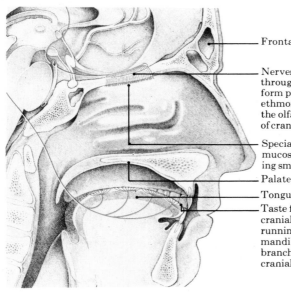

Frontal sinus

Nerves passing through the cribriform plate of the ethmoid bone to the olfactory bulb of cranial nerve (I).

Specialized nasal mucosa for detecting smell.

Palate

Tongue

Taste fibres from cranial nerve (VII) running in the mandibular branch of the cranial nerve (V).

Referred pain

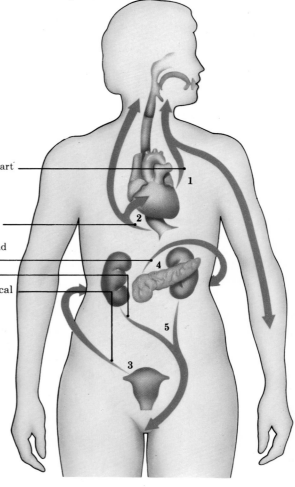

From the heart

From the esophagus

Pancreas and stomach

Kidney

Gynecological

Smell

The olfactory nerves start in a small patch of special nasal mucosa high in the nasal passages (*above*). They pass through minute holes in the cribriform plate of the ethmoid bone into the olfactory bulb of the cranial nerve (I) before entering directly into the limbic system of the brain.

It is thought that there are several types of chemoreceptors and that odours are differentiated by the amount of stimulus each is given. Like taste cells, these chemoreceptors have hair-like microvilli on the surface. The dissolved, odiferous molecules may fit into some, but not into all, types of cell microvilli, causing an electrical discharge and stimulus of the nerve. It is thought that some substances, pheromones, act as sexual stimulants. Well-trained sense receptors can recognize over 10,000 different odours.

Taste—the tongue

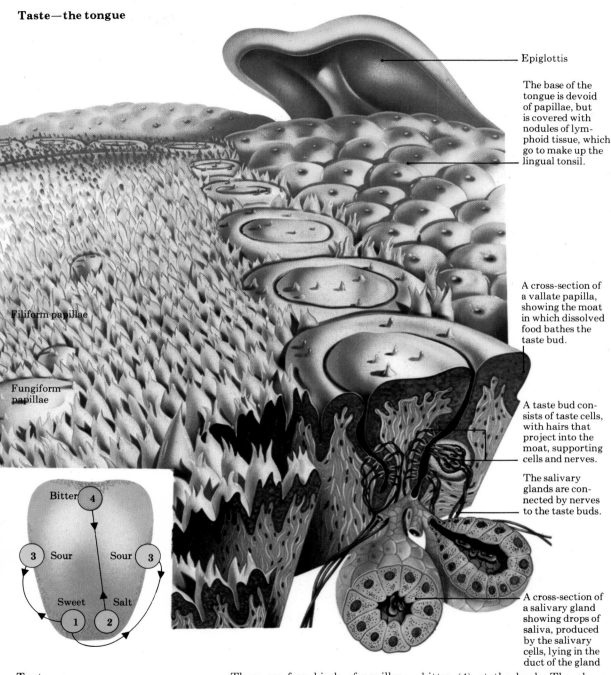

Epiglottis

The base of the tongue is devoid of papillae, but is covered with nodules of lymphoid tissue, which go to make up the lingual tonsil.

Filiform papillae

Fungiform papillae

A cross-section of a vallate papilla, showing the moat in which dissolved food bathes the taste bud.

A taste bud consists of taste cells, with hairs that project into the moat, supporting cells and nerves.

The salivary glands are connected by nerves to the taste buds.

A cross-section of a salivary gland showing drops of saliva, produced by the salivary cells, lying in the duct of the gland.

Bitter 4
3 Sour | Sour 3
Sweet | Salt
1 | 2

Taste

The highly mobile muscular tongue (*above*), supplied by the hypoglossal nerve (XII), is attached to the hyoid bone and membrane between the mandibles. It aids mastication and articulation of speech. The surface is covered by thick epithelium containing about 9,000 papillae—taste buds—which are re-formed and replaced within 48 hours, but become fewer with age.

There are four kinds of papillae: Filiform, folioform and vallate, which are only found at the back of the tongue, and fungiform. Each contains one to two hundred microvilli covered with chemoreceptors. A single nerve may connect several cells.

All flavours depend on a mixture of four basic tastes (*see* inset): Sweet (1) and salt (2), mainly on the front, and sour (3), at the sides, and bitter (4) at the back. The glossopharangeal nerve (IX) carries sensation from the posterior third of the tongue. Taste nerves in the mandibular branch of the nerve (V), supplying the rest of the tongue, branch to join the facial nerve (VII).

Palatability is a combination of taste and temperature, feel and smell. It is assessed by the parietal lobe of the brain in conjunction with the thalamic taste centre.

Definitions

Chemoreceptor Sensory nerve ending responsive to chemical stimulation.

Cribriform plate Part of the ethmoid bone.

Ethmoid bone Forms part of the skull adjacent to the nasal passages.

Hyoid bone U-shaped bone above the larynx and surrounding the anterior part of the pharynx.

Limbic system Pair of structures above the thalamus, concerned with memory and instinctive emotions.

Olfactory lobe Swelling at the end of the olfactory nerve.

Papilla Small cone-shaped elevation; filiform is hair-like, foliate leaf-like, fungiform mushroom-like, vallate trench-like.

Parietal cortex Grey matter in the brain concerned with sensation.

Phrenic nerve Arises from the third, fourth and fifth cervical nerves and supplies the diaphragm.

Taste centre Pair of nuclei lying in the thalamus.

Thalamus Grey matter at the brain's base concerned with sensory nervous transmission.

Common diseases

Anosmia Loss of sense of smell.

Cancer of the tongue May occur owing to chronic irritation, e.g. smoking.

Glossitis Inflammation of the tongue.

Leukoplakia Chronic condition of the tongue.

Cross-references

Digestion 44
Nervous System 57–61
Skin 70

Sex 1

Male anatomy

Much of the male reproductive anatomy is external. The two testes hang in the scrotal sacs surrounded by the tunica vaginalis. The vasa efferentia join the testis to the overlying epididymis and vas deferens, which join the urethra in the centre of the prostate gland. Akin to the gall bladder the seminal vesicle acts as a storage organ for the mature sperm. It lies between the prostate gland and the colon.

The wrinkled subcutaneous area of the scrotum contains the dartos muscle, which can contract and, with the cremaster muscle, attached to the spermatic cord, pull the testes closer to the abdomen. The spermatic cord is the combination of the vas deferens (surrounding testicular blood vessels and nerves), fatty tissue and the projection from the peritoneum that originally formed the tunica vaginalis.

The prostate gland lies around the first part of the urethra at the base of the bladder, and its secretions help maintain sperm activity. Two additional pairs of glands, Cowper's and Littre's, empty into the urethra adjacent to the prostate. They produce the bulk of the ejaculate of 3 to 5 millilitres of semen.

Semen contains various proteins, fructose and a mixture of chemicals, which help with the nutrition of sperm in the vagina. Each millilitre contains about 100 million sperm, of which 20 per cent are dead or abnormal. The number of sperm produced at ejaculation is variable.

The penis surrounds the fibro-elastic urethra, which runs in the corpus spongiosum and ends in the glans, which is covered by the foreskin. The urethra and corpus spongiosum are covered, above and at the sides, by the pair of corpora cavernosa. Normally the penis is limp, but if the muscles at the base contract around the veins the spongy tissue in all three corpora becomes congested with blood and an erection occurs. The penis has two functions: The excretion of urine from the bladder out of the body; and the deposition of semen in the vagina of the female genital tract.

Male reproductive organs

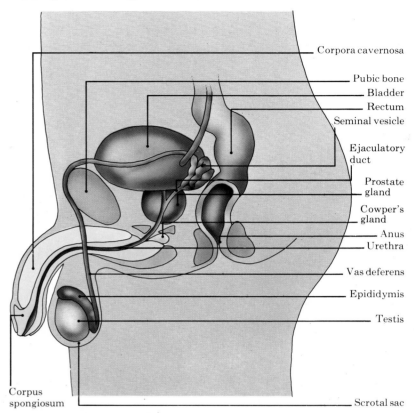

Corpora cavernosa
Pubic bone
Bladder
Rectum
Seminal vesicle
Ejaculatory duct
Prostate gland
Cowper's gland
Anus
Urethra
Vas deferens
Epididymis
Testis
Corpus spongiosum
Scrotal sac

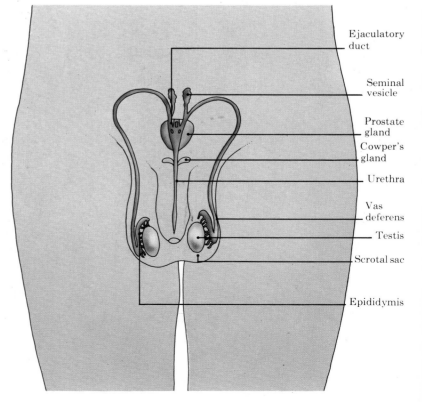

Ejaculatory duct
Seminal vesicle
Prostate gland
Cowper's gland
Urethra
Vas deferens
Testis
Scrotal sac
Epididymis

Female reproductive organs

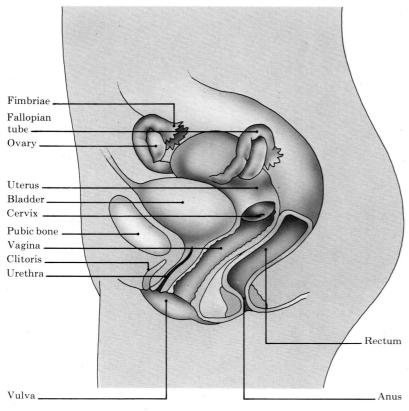

Fimbriae
Fallopian tube
Ovary

Uterus
Bladder
Cervix
Pubic bone
Vagina
Clitoris
Urethra

Rectum

Vulva
Anus

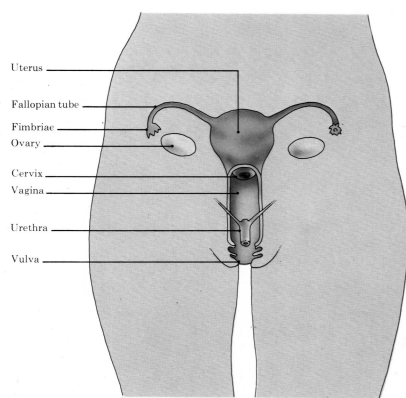

Uterus

Fallopian tube

Fimbriae

Ovary

Cervix

Vagina

Urethra

Vulva

Female anatomy

The female reproductive system not only has to produce an ovum but has to nurture the fertilized ovum and protect it until pregnancy ends. At the entrance to the vagina a pair of lip-like folds—the larger, thicker labia majora and the smaller, inner labia minora—lie along either side of the vaginal entrance and join in the front, blending into the padded, hairy area of the mons pubis. At the front they enclose the exit of the urethra just behind the small projection of erectile tissue—the clitoris—comparable to the penis.

Behind these structures is the vagina—a 10 to 15 centimetre elastic tube lined with moist epithelium; it is normally "closed" except during intercourse. At the back of the vulva are two large Bartholin's glands, and lining the entrance to the vagina a large number of smaller, lubricating glands. At the top of the vagina the uterus is held in place by the muscles and four strong fibrous ligaments of the pelvic floor, and to the side of the pelvis by pairs of the round and suspensory ligaments, running in folds of peritoneum.

The uterus is a small pear-shaped organ, covered with peritoneum, with a thick wall of interweaving muscle fibres and lined with the special endometrial cells. It is situated behind the bladder and in front of the rectum. The cervix of the uterus is a thick fibro-muscular structure opening into the vagina and lined with special cells that form a "plug" of mucus. The uterine muscle is always contracting and relaxing slightly. This is increased during sexual orgasm to "suck in" sperm; during menstruation, to expel the endometrium; and also at parturition.

The two Fallopian tubes are about 10 centimetres long with finger-like fimbriae at the ends to encircle the ovaries. The ovum is swept down the tube by a combination of ciliated epithelium and peristaltic muscular contraction. Unlike the male urinary system, that of the female is separate from the reproductive system. The bladder empties into the urethra, which opens in front of the vagina.

Endometrium Hormone-responsive layers of cells that line the uterine cavity.

Epididymis Tightly coiled 6 m tube joined to the upper part of the testis by 12 to 20 vasa differentia, in which the sperm spend 10 days maturing.

Fallopian tube Two 10 cm tubes extending from the fimbriae, surrounding the ovaries, that join the body of the uterus.

Fimbria One of the soft finger-like processes at the peripheral end of the Fallopian tube.

Follicular cyst Ovarian cyst that develops each month and from which the mature ovum bursts into the peritoneal cavity.

Glans penis Swelling at the distal end of the corpus spongiosum, which is covered by the foreskin in the uncircumcised male.

Graafian follicle *See* Follicular cyst.

Implantation Process by which the embryo invades and erodes the endometrial lining of the uterus.

Labia majora Two larger and outer folds of skin that surround the entrance to the vagina.

Labia minora Two smaller and inner folds of skin that surround the entrance to the vagina.

Lacuna Small areas in the endometrium, formed by the trophoblast, in which the maternal blood bathes the trophoblastic cells.

Littre's gland One of a pair of glands adjacent to the prostate whose secretions help maintain the health of the sperm in the ejaculate.

Mitochondria Minute structures within the cell that store the energy required for cell metabolism.

Sex 2

The testis

The testis (right) has two functions: The production of testosterone; and of spermatozoa. It is about 5 centimetres long and 2.5 centimetres thick and surrounded by the tunica albuginea; this coat is divided into about two hundred lobes containing four to six hundred seminiferous tubules, each about 75 centimetres long. The tubules produce about 200 million sperm every day which pass into a series of communicating ducts —the rete testis. These ducts, lined with ciliated cells, join to form 12 to 20 vasa efferentia, which run to the epididymis, where the sperm spend up to ten days maturing before entering the vas deferens. The sperm can then survive a further six weeks before degeneration and absorption.

Spermatogenesis requires stimulus from the follicular stimulating hormone of the anterior pituitary lobe. Testosterone responds to luteinizing hormone. It causes development of male secondary sexual characteristics—pubic and facial hair growth, aggressiveness, muscle bulk and deepening of the voice.

The testis

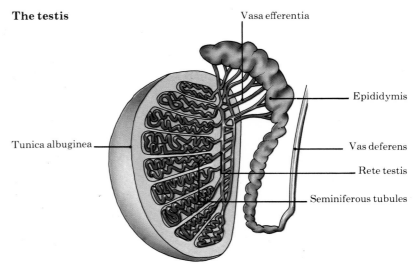

Vasa efferentia

Epididymis

Vas deferens

Rete testis

Seminiferous tubules

Tunica albuginea

Spermatogenesis

The spermatogonia divides (below) by mitosis to form the primary spermatocyte, containing 46 chromosomes. Meiosis occurs when half of each chromosome pair enters the secondary spermatocyte. This divides again to produce a spermatid that finally matures into a spermatozoon.

The spermatozoa

The mature sperm (below) is 0.05 millimetres long. It consists of a head, body and tail. The head is covered by the acrosome cap and contains a nucleus of dense genetic material from the 23 chromosomes. It is attached by a neck to a body containing mitochondria that supply the energy for the sperm's activity. The tail is made of protein fibres that contract on alternate sides, giving a characteristic wave-like movement that drives the sperm through the seminal fluid, which also supplies additional energy. Some sperm have two heads or two tails and if the testes are too warm they may die or spermatogenesis may not occur.

Spermatogenesis

Spermatogonia

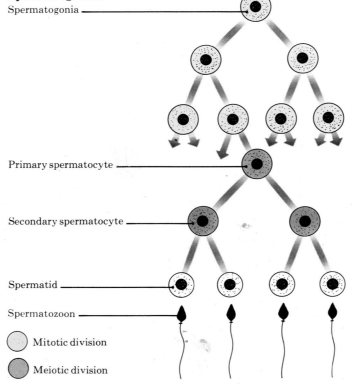

Primary spermatocyte

Secondary spermatocyte

Spermatid

Spermatozoon

○ Mitotic division

● Meiotic division

Spermatozoa

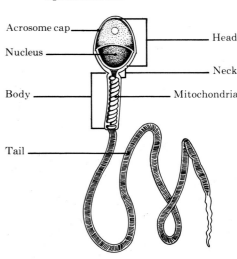

Acrosome cap

Nucleus

Body

Tail

Head

Neck

Mitochondria

The ovary

The ovary has two functions: The production of ova (*right*); and the secretion of estrogen and progesterone. It is about 2 centimetres across and 1 centimetre thick.

At puberty, the onset of menstruation and the development of the secondary sexual characteristics—hair growth, breast development and redistribution of fat to the buttocks and shoulders—are the result of the effect on the ovaries of the increasing pituitary secretion of follicular stimulating hormones (FSH) and luteinizing hormones (LH).

Like spermatogenesis an oocyte is formed from the oogonia. Meiotic division then occurs to produce a large secondary oocyte with most of the cytoplasm from its "twin"—the polar body. Further mitotic division occurs so that one primary oocyte has produced a large ovum and three polar bodies, which finally disintegrate in the ovum.

The ovary has 50 to 250,000 oogonia, but only about 500 eventually become mature ova.

The menstrual cycle

From the menarche to the menopause the anterior pituitary lobe maintains a rhythm of FSH and LH secretion to produce, in most women, a regular menstrual cycle (*right*). FSH stimulates the maturation of several Graafian follicles, of which only one reaches maturity. These follicular cells produce estrogen to build up the endometrium. In mid-cycle a surge of LH softens the plug of cervical mucus and causes ovulation. The corpus luteum is formed and secretes progesterone, which, with the estrogen, further prepares the endometrium for implantation of the fertilized egg.

If fertilization occurs the embryo produces human chorionic gonadotrophin to maintain the stimulus to the corpus luteum and continued production of progesterone. If fertilization does not occur the corpus luteum degenerates, progesterone production ceases and the endometrium is shed causing menstruation for about five days. The change in hormonal balance may be responsible for premenstrual tension.

The ovary

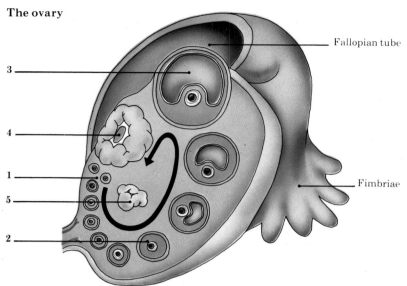

Fallopian tube

Fimbriae

The ovary

The cross-section of the ovary (*above*) shows an immature ovum (1) absorbing fluid and swelling into the sac-like Graafian follicle (2). As it reaches the surface (3), half-way through the menstrual cycle, the ovum bursts out into the peritoneal cavity to be collected by the fimbriae. The empty follicular cyst then changes into the corpus luteum (4), which grows until menstruation, when it shrivels into a scab-like corpus albicans (5).

The menstrual cycle

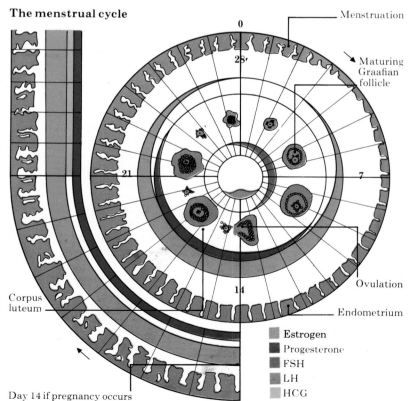

Menstruation

Maturing Graafian follicle

Ovulation

Endometrium

Corpus luteum

Day 14 if pregnancy occurs

Estrogen
Progesterone
FSH
LH
HCG

Sex 3

Coitus

Coitus is the insertion of the erect penis into the vagina followed by rhythmical movements ending in orgasm and ejaculation of semen by the man. Initially sexual excitement is produced by sight, touch and sound and, perhaps, pheromones that stimulate smell. In the man there is congestion of the corpora cavernosa and corpus spongiosum of the penis; in the woman there is slight engorgement of the breasts, and congestion of the clitoris and labia, with increased vaginal secretions. A prolonged phase of enjoyment can then follow when the penis is inserted into the vagina provided the "crescendo" to orgasm is controlled.

Orgasm is a series of rapid muscular contractions surrounding the male urethra with ejaculation of the semen. This can be complemented by a simultaneous happening in the woman in which the upper part of the vagina and a rhythmical "sucking in" of the uterus draw the sperm to the right area. During sexual intercourse there are increases in blood pressure, rates of heartbeat and respiration. After orgasm there is physical relaxation.

Fertilization

The moment of conception is the most important stage of sexual reproduction. The joining of the two nuclei, each containing 23 chromosomes, to form a cell of 46 chromosomes, will produce an embryo.

The sex of the baby is decided by the father's chromosomes—the XY sex pair. When the primary spermatocyte divides by meiosis to form the secondary spermatocytes, containing 23 chromosomes, either an X or a Y chromosome will move into each cell. The female pair of chromosomes are the same, XX, so all ova will contain the X chromosome. The fertilizing sperm will join the ova to form an XY male or XX female embryo. It is probable that there is less than 24 hours in which the ovum can be fertilized and as the sperm survive for about 48 hours in the uterus and tubes, there is only a limited time each month in which conception can occur.

Sperm transport

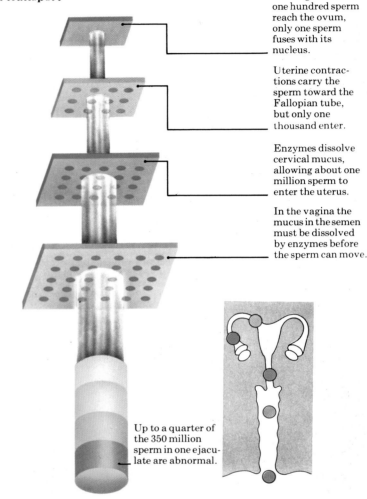

Although about one hundred sperm reach the ovum, only one sperm fuses with its nucleus.

Uterine contractions carry the sperm toward the Fallopian tube, but only one thousand enter.

Enzymes dissolve cervical mucus, allowing about one million sperm to enter the uterus.

In the vagina the mucus in the semen must be dissolved by enzymes before the sperm can move.

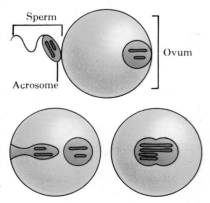

Up to a quarter of the 350 million sperm in one ejaculate are abnormal.

The sperm's journey to the ovum

The ejaculate of semen contains more than 350 million sperm. Ideally they are deposited adjacent to the cervix, where the vaginal enzymes dissolve the seminal mucus and release the sperm. They then swim into the cervical mucus, which is watery and soft enough at ovulation to allow many of the sperm to penetrate to the uterine cavity. Uterine contractions will then help them to move upwards to the Fallopian tubes, where they again swim, perhaps with the help of the lining of ciliated cells, to reach the ovum at the mid-point of the tube. Only about 100 sperm survive the journey of nearly 24 hours and only one fertilizes the ovum. The ultimate fate of the surplus spermatozoa in the female genital tract is still not known.

The moment of conception

Sperm

Acrosome

Ovum

The sperm's acrosome disappears as it dissolves the membrane of the ovum (*above*). The tail and body are shed when the head penetrates to join its 23 chromosomes with those of the ovarian nucleus.

Implantation of the ovum

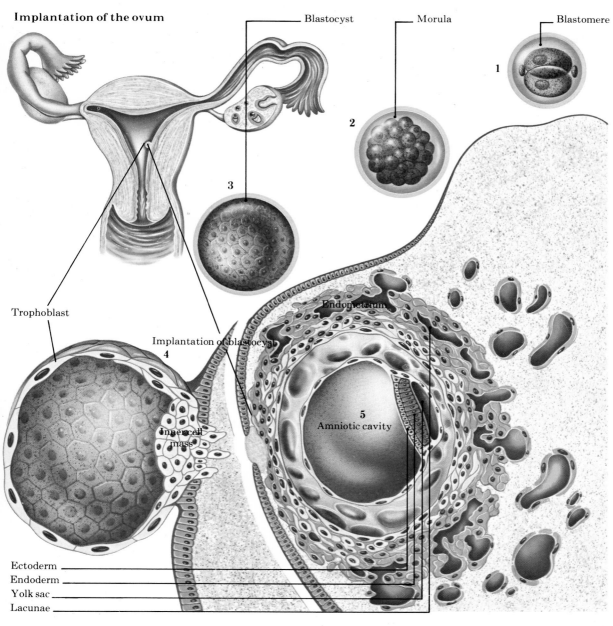

Blastocyst — Morula — Blastomere

1

2

3

Trophoblast

Implantation of blastocyst
4

Endometrium

Inner cell mass

5
Amniotic cavity

Ectoderm
Endoderm
Yolk sac
Lacunae

Implantation of the ovum

It takes about a week for the fertilized ovum to pass down the Fallopian tube and implant itself in the endometrium. Within hours of conception mitosis begins with the development of a sphere of an increasing number of cells; the sphere starts as the blastomere (1), then becomes a morula (2) of about 64 cells. At this stage it changes into a hollow, fluid-containing ball —blastocyst (3)—with the inner cell mass at one end. It can now begin implantation (4). The outer ring of cells, now called the trophoblast, secretes enzymes that erode the endometrium. The trophoblastic cells spread into the endometrium forming lacunae—fluid-containing sacs—that penetrate the maternal circulation. This allows nutrition to take place. Ultimately the trophoblast forms the outer layer of the placenta.

The inner cell mass splits into two layers: Endoderm—ultimately the alimentary tract—which produces a yolk sac; and the ectoderm —ultimately the skin, brain and spinal cord—which produces the amniotic sac to surround and protect the embryo in a bag of fluid.

By the ninth day after conception the blastocyst has sunk deep into the endometrium (5) and is already receiving nutrition from the mother. Human chorionic gonadotrophin is secreted by the trophoblast to maintain the progesterone production from the corpus luteum. This will continue until the end of the third month, when the placenta produces sufficient hormone to maintain this function until the end of pregnancy.

Cross-references

Pregnancy 1

Development of cell layers

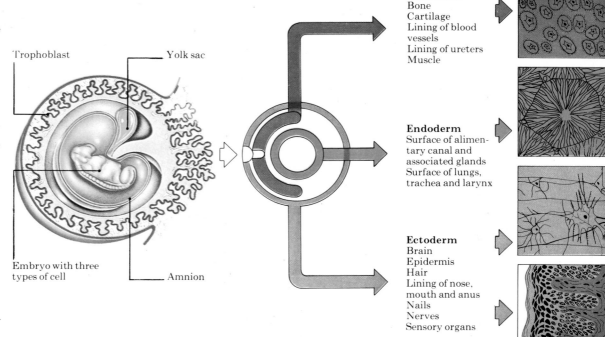

Trophoblast

Yolk sac

Embryo with three types of cell

Amnion

Mesoderm
Bone
Cartilage
Lining of blood vessels
Lining of ureters
Muscle

Endoderm
Surface of alimentary canal and associated glands
Surface of lungs, trachea and larynx

Ectoderm
Brain
Epidermis
Hair
Lining of nose, mouth and anus
Nails
Nerves
Sensory organs

The first four weeks

The blastocyst begins to implant into the endometrium on the seventh day after conception. It is composed of two distinct parts, a covering layer of the trophoblastic cells and an inner cell mass.

The trophoblast invades the maternal tissues and erodes the maternal blood vessels so that the surface is bathed in oxygen-carrying nutritious fluid. In addition the trophoblastic cells produce human chorionic gonadotrophin to stimulate the production of progesterone and maintain pregnancy.

The inner cell mass develops a fluid-containing cavity from its upper surface—the amniotic sac—which eventually expands to line the inner surface of the trophoblastic cell layer. This layer ultimately forms the chorion of the fetal membranes, and protects the embryo and fetus until parturition.

Three cell-producing layers

By the sixteenth day the inner cell mass has two layers of cells (*above*): The ectoderm, underlying the amniotic cavity, and the endoderm, which starts to form another fluid-containing cavity—the yolk sac. At about the same time the two layers start to develop a stalk, the beginnings of the umbilical cord, separating them from the trophoblastic layer.

At first the embryo receives nutrition from the trophoblast and the yolk sac, but gradually the thick trophoblastic layer, outside the area of implantation, regresses, leaving the implanted part to form a thick disc of tissue—the placenta. The stalk to the embryo grows blood vessels and incorporates the yolk sac as it becomes surrounded by the amniotic membrane.

In the third week the two layers of the inner cell mass produce a third layer sandwiched between them—the mesoderm (*above*).

The endoderm eventually forms the intestines, liver, gall bladder and pancreas. It also forms the larynx and trachea from a pouch at the upper end, and finally the lungs from small buds.

The mesoderm is the precursor of all the muscles, bones and blood vessels. The ectoderm supplies the surface of the body—skin, hair and nails; the lining of the nose, mouth and anus; and the nervous system.

From the third week the most rapid development of the embryo takes place. It is transforming from a flat disc of cells surrounded by fluid into a recognizable shape. The heart begins to develop as the basic tube of the primitive blood vessel divides and twists. The nervous system develops from a straight groove appearing along the middle of the ectoderm. The two edges curl towards each other and join to form the neural tube—the primitive spinal cord—and, at one end, a bulge appears that ultimately changes into the brain. The central channel remains, becoming the central canal of the spinal cord and ventricles of the brain, in which cerebrospinal fluid will flow.

This division into three layers is a prelude to the rapid changes of the next four weeks, when all the basic organs and structures are formed.

The next two months

By the end of the third week (*below right*) the mesodermal tissue is arranging itself into somites—a segmental arrangement that can be seen in humans in the ribs of the thoracic cage. Each somite of muscle is eventually supplied by its own blood vessels and nerves.

At four weeks the embryo is 3 millimetres long. It consists of 25 somites, a large bulge of the heart, and has small ear pits in the head. It is not until the fifth week that a pair of pigmented discs appear on the head. These are the first signs of the eyes. In front is the small depression of the primitive nose. The limb buds, arms first and then legs, begin to grow by the sixth week.

At six weeks the embryo is just recognizable as a "being", with pulsating heart, forty somites and tiny ear flaps. At this stage it is nearly 1.5 centimetres long and the hands, fingers and feet are just recognizable.

Next the somites lose their well-defined divisions and join up to form a solid mass of body tissue with recognizable thorax, abdomen and the beginnings of a neck. By the seventh week the bulbous protuberance is clearly a head with well-defined nose, ears, eyes and mouth. The pelvis is starting to form and the end somites lose their tail-like appearance as the sacral bone is developed.

It is during this rapidly changing period, from the fourth to the seventh week, that congenital malformations may occur if the tissues fail to fold or divide correctly. External influences, such as disease or drugs, may damage the embryo.

At the eighth week the embryo is 2.5 centimetres long and recognizably human, with eyelids and separate fingers and toes developing.

From the end of the second month onwards development is mainly by growth and some alteration in proportion, so by the end of twelve weeks the embryo is 5 centimetres long and the mother's uterus is the size of an orange. It is at this stage of development that the embryo is called a fetus.

Growth of embryo

Actual size

Three-week embryo Four-week embryo Five-week embryo

Six-week embryo Seven-week embryo Eight-week fetus

Maintaining pregnancy

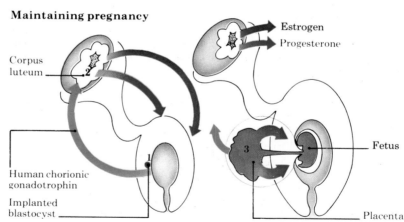

Estrogen
Progesterone

Corpus luteum

Human chorionic gonadotrophin
Implanted blastocyst

Fetus

Placenta

Hormonal control of pregnancy

At the moment of conception the corpus luteum is producing progesterone and estrogen to prepare the endometrium for implantation (*above*). Before the corpus luteum starts to degenerate, the trophoblast layer of the blastocyst begins to secrete human chorionic gonadotrophin. This starts about the fifth day and rapidly increases after implantation (1) to maintain the luteal secretions (2). This maintains the condition of the endometrium and suppresses further ovulation.

By the twelfth week the placenta produces more human chorionic gonadotrophin and also its own hormones (3) to maintain pregnancy. Until this stage the corpus luteum has grown in size, but it now begins to degenerate. If the increase in placental hormones does not keep pace with the reduction of those from the corpus luteum the placenta may detach and abortion occur.

Ductus venosus Fetal vein running from the umbilical vein to the inferior vena cava which bypasses the hepatic circulation obliterated at birth.

Embryo Developing fetus from the moment of conception to the end of the third month.

"Engage" Moment at which the fetal-presenting part enters the pelvis. This is considered to be the lower part, where the area is completely surrounded by bone.

Ergometrine Drug used to help with the contraction of the uterus following parturition.

Estrogen Ovarian hormone, which helps to prepare the endometrium for implantation of the fertilized ovum.

"Face to pubis" When the baby is born with the occiput in the posterior position.

Fetus Developing infant from the end of the third month of intra-uterine life to parturition.

Foramen ovale Opening between the two atria in the fetal heart. It closes at birth.

Hemolysis Breakdown of red blood cells.

Human chorionic gonadotrophin Hormone secreted by the trophoblast to continue stimulation of the production of progesterone from the corpus luteum.

Implantation Method by which the blastocyst penetrates and adheres to the endometrium.

Inner cell mass Collection of cells in the blastocyst that forms the developing embryo.

Labour Childbirth. It begins with the onset of regular uterine contractions and ends with the expulsion of the baby and placenta.

Pregnancy 2

Lactation Production of milk by the mammary glands.

Lanugo Fine hair that is found on the fetal body.

"Lightening" Sensation of increased abdominal comfort when the fetal-presenting part engages, thus giving more room in the abdomen.

Lochia Vaginal discharge that occurs after parturition.

Mesoderm Middle layer of the inner cell mass that ultimately forms the muscles, bone and cardio-vascular system.

Mitosis Division of a cell to produce an identical one containing the same number of chromosomes.

Neural tube Precursor of the central nervous system which is formed by the ectoderm.

Occiput Back of the head.

Ossification centre Area in tissue from which bone is formed.

Ovulation Moment when an ovum escapes from the ovary.

Oxytocin Hormone produced by the posterior lobe of the pituitary gland that, with estrogen, helps to initiate and maintain uterine contractions in labour.

Parturition Process of giving birth to a child.

Placenta Disc-like organ, about 25 cm in diameter, 3 cm thick and weighing about 600 g, composed of about 50 cotyledons and connected to the foetus by the umbilical cord. The placental function is to exchange fetal carbon dioxide and waste metabolites for oxygen and nutrition. It is able to excrete metabolites. It also secretes human chorionic gonadotrophin in the first three months of pregnancy and progesterone and estrogen.

Five-month fetus

Nine-month fetus

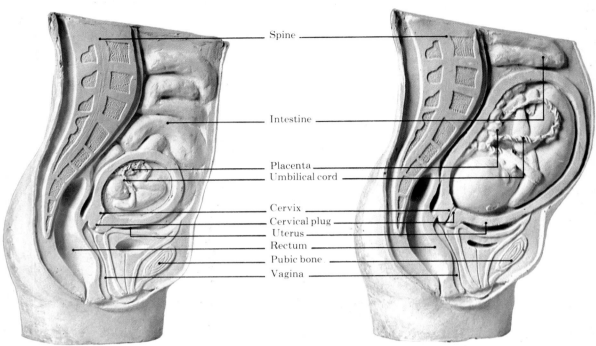

- Spine
- Intestine
- Placenta
- Umbilical cord
- Cervix
- Cervical plug
- Uterus
- Rectum
- Pubic bone
- Vagina

The continuation of pregnancy

At the end of the fourth month the fetus weighs about 125 grammes and is 12.5 centimetres long, the genitalia are developing and ossification centres appear. It is moving in the amniotic fluid, and in the fifth month (*above left*) the mother may feel these movements—"quickening"—for the first time.

In the sixth month the fetus starts to produce a layer of fat under the skin and a fine hair—lanugo—covers the head and body, which is also covered by a greasy substance—vernix—that protects and lubricates the fetal skin, particularly during parturition. As the fetus grows the proportion of amniotic fluid decreases. At 28 weeks the fetus

weighs about 500 grammes and the heart-beat of 140 a minute can easily be heard; the fetus is now viable. Four to six weeks later the head engages in the pelvis and remains there until parturition. During the last two months (*above right*) it matures steadily and gains about 25 grammes weight a day to reach about 3 kilogrammes at birth.

Fetal sex

Fetal sex depends on the male Y chromosome and fetal secretion of small amounts of testosterone at about the fourteenth week of pregnancy. About 106 boys are born for every 100 girls. However, there are fewer adult males owing to a greater male infant mortality.

Twins

Twin births (*left*) occur about once in every 85 pregnancies. Identical twins (**A**) are always the same sex, developing from mitotic division of the same fertilized ovum. Each fetus has its own amniotic sac, but shares a single placenta. Fraternal, non-identical, twins (**B**), develop from two separate ova that have been fertilized at the same time by different spermatozoa. They have separate placentae

Twin births

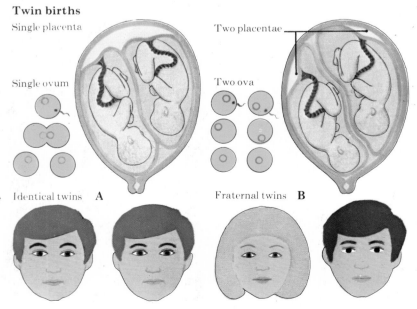

Single placenta

Single ovum

Two placentae

Two ova

Identical twins **A**

Fraternal twins **B**

Circulation in the fetus and at birth

Throughout pregnancy oxygen reaches the fetus (*right*) from the placenta via the umbilical vein. Most of the blood passes through the liver. Some passes into the ductus venosus (3) and into the general circulation, after it has passed through the liver, to reach the heart. By this time it is only partially oxygenated.

In the heart much of the blood passes through the foramen ovale (1) into the left atrium and back into the systemic circulation. Most of the blood from the right ventricle is pumped through the pulmonary artery and a special bypass artery, the ductus arteriosus (2), to the systemic vessels. Only a small amount continues through the pulmonary system. Blood is then pumped through the systemic system and the two umbilical arteries to reach the placenta.

At birth (*far right*) the placental circulation ceases, the lungs expand, causing the ductus arteriosus and foramen ovale to close, and the circulation becomes independent.

The mother in pregnancy

In the first trimester the hormonal changes may cause morning nausea and vomiting with fatigue, but there is little weight gain. After this period these unpleasant symptoms disappear and there is a steady gain in weight. This is only partly due to the growing fetus. The uterus, amniotic fluid and placenta contribute a certain amount. Over half the weight gain is a combination of fluid retention, which also softens joints and ligaments, and reserves of fats for the energy required during lactation. The mother will notice the "quickening" soon after the sixteenth week and then increasing awareness of movement and kicking until the end of pregnancy. There is often increased comfort—"lightening"—about the thirty-sixth week, when the head engages. This is due to a lowering of the uterus from the xiphisternum.

The mother's diet requires increased proteins, calories, vitamins, iron and calcium. Regular exercise and rest contribute to good health.

Circulation before birth

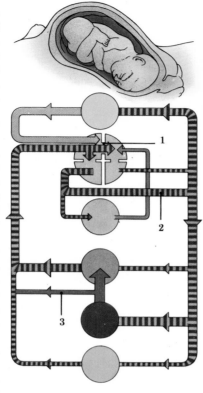

⬤	Head and arms
⬤	Heart
⬤	Lungs
⬤	Liver

The placenta

The placenta is fully developed by mid-pregnancy with 50 to 60 cotyledons each supplied by a branch of the umbilical artery. It reaches an ultimate diameter of about 25 centimetres with a surface area of 10 square metres. The umbilical cord of two arteries, one vein and Wharton's jelly is usually attached to the centre. The placental function is to exchange the fetal carbon dioxide and waste metabolites for oxygen and nutrition.

The placenta is bathed in maternal blood brought by spinal arteries through the uterine wall. The two circulations are separate, but leaks of fetal cells into the maternal blood may cause problems in Rhesus blood group incompatibility.

Circulation after birth

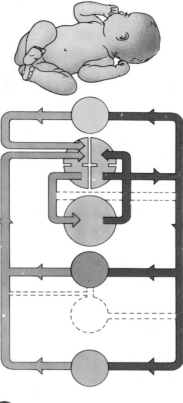

⬤	Placenta
⬤	Body
▬	Oxygenated blood
▬	Deoxygenated blood
▥	Mixed blood containing oxygen and waste materials

The placenta—maternal and fetal blood vessels

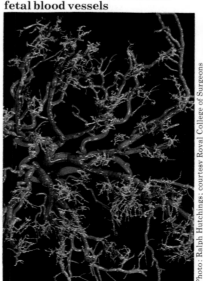

Photo: Ralph Hutchings; courtesy Royal College of Surgeons

Pregnancy 3

Sequence of normal birth

Abortion Loss of the fetus before the twenty-eighth week of pregnancy.
Criminal—Termination that is illegally produced.
Incomplete—When only part of the fetus or placenta has been expelled.
Inevitable—Regular pain and bleeding indicate that an abortion will take place.
Missed—Death of the fetus or embryo without expulsion. This can lead to formation of a carneous mole.
Septic—Abortion in the presence of uterine infection.
Therapeutic—Termination of pregnancy for legally acceptable reasons.
Threatened—Bleeding without pain.

Antepartum hemorrhage
Accidental—Bleeding from a normally situated placenta.
Extra-uterine—Bleeding from outside the uterus, e.g. cervix.
Unavoidable—Bleeding from a placenta previa.

Chorion-epithelioma Malignancy occurring in a hydatidiform mole.

Chromosome abnormalities
Many abnormalities of this kind may abort, but some, e.g. Down's syndrome, may survive.

Ectopic pregnancy
Implantation in some place other than the uterus, e.g. Fallopian tube or ovary. This may occur if there has been a delay in the movement of the fertilized ovum towards the uterus.

Glycosuria This may occur owing to diabetes mellitus appearing under the stress of pregnancy or to a change in the kidneys' ability to reabsorb glucose.

Hemorrhoids
Can occur during pregnancy due to venous congestion below the presenting part and swelling of rectal and pelvic tissues.

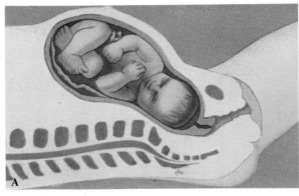

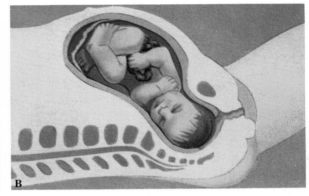

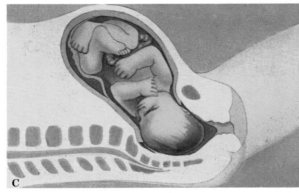

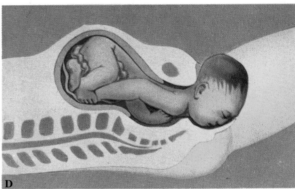

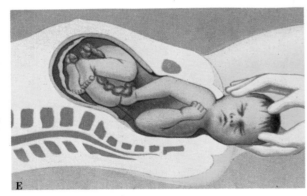

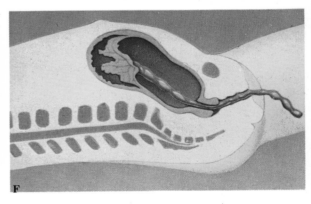

Childbirth

At about 280 days after conception the fetus is expelled from the uterus (*above*). If pregnancy continues for too long the placenta begins to degenerate and the fetus is slowly starved of oxygen and nutrition.

Throughout pregnancy the uterine muscle has been contracting and relaxing. During the last few weeks these contractions have become stronger and are sometimes noticed by the mother. They help to keep the fetal head engaged in the pelvis, (**A**). Labour is divided into three stages: The first is from the onset of labour, with regular rhythmical uterine contractions, to the moment when the cervix is fully dilated. This may last from two to twenty-four hours or sometimes longer. The second stage is from the full dilatation of the cervix to the moment of delivery and lasts less than two hours. The third stage is the expulsion of the placenta, cord and membranes. The placenta has to be carefully examined to make sure that parts of it have not been left behind which could cause a hemorrhage

The first stage

The first stage starts with backache and regular uterine contractions that become gradually stronger and by the end occur about every two minutes, each lasting thirty to forty seconds. As the head is forced deeper into the pelvis it turns sideways (**B**) and the plug of cervical mucus is dislodged causing a "show" of blood. The cervix is gradually stretched, thinned and dilated until it is wide enough (**C**) for the head to pass into the vagina. If it has not already done so, the amniotic sac ruptures.

Breech birth

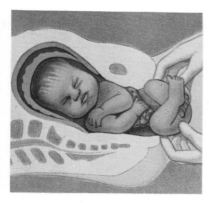

The second and third stages

The second stage starts with the cervix fully dilated (**C**). The head moves deeper into the pelvis and again turns so that the occiput is behind the symphysis pubis. The mother has a "bearing down" sensation and can help the powerful uterine contractions to force the fetus through the vagina. The vulva distend around the head and finally it is "crowned" when the greatest circumference is reached. The head extends (**D**) round the symphysis as it is pushed out. The baby's first breath is taken and is followed by a cry. This expands the lungs and starts respiration, as well as obliterating the fetal circulation.

The baby can now be delivered (**E**) with the next contraction as the smaller shoulders and body easily follow the large head. At this moment the mother is usually given an injection of oxytocin and ergometrine to stimulate uterine contraction and the separation of the placenta.

The third stage follows ten to twenty minutes later when the placenta has separated from the uterus and is expelled (**F**) along with the umbilical cord and membranes.

A breech baby

A breech presentation (*above*) is when the buttocks engage in the pelvis instead of the head and delivery is complicated as the largest part, the head, is delivered last. Breathing cannot start until the same time as the placenta begins to separate with the obvious hazards of oxygen deprivation.

The puerperium

After parturition the uterus returns to normal size in about a month with loss of lochia, which is initially fresh blood later becoming creamy in colour. Menstruation will return in about two months unless suppressed by prolactin secretion during lactation. Excess body fluid and fat gradually disappear and softened ligaments regain their strength.

The baby is born with excess fluid, which it loses in the first few days, and a raised hemoglobin count level, which falls slowly with the appearance of slight jaundice. This is the result of raised levels of serum bilirubin from hemolysis and the immature liver's inability to excrete it.

Lactation

Throughout pregnancy the breasts have been enlarging under the stimulus of the anterior pituitary hormone, prolactin, to produce a watery milk (*below*). In the last few weeks before parturition colostrum is produced and secreted for the first few days. Then normal milk (1) is produced and the infant's sucking (2) stimulates the hypothalamus. This continues the production of prolactin and also oxytocin from the posterior pituitary lobe (3), which causes the breast alveoli to contract and force—"let down"—the milk into the ducts (4).

Hormones in labour

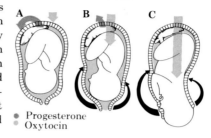

● Progesterone
● Oxytocin

Hormones in labour

The story of hormones in labour (*above*) is not fully understood. In some way their interreaction adjusts the actual onset of labour. Throughout pregnancy the progesterone has made the uterine muscle (**A**) relaxed and unresponsive to the posterior pituitary hormone—oxytocin. Towards the end of pregnancy the progesterone production falls and there seems to be an increase in estrogen. The combination of oxytocin and estrogen initiates uterine contractions. The dilatation of the cervix (**B**) increases oxytocin output and further reduces progesterone, so the contractions become stronger, longer and more frequent (**C**). In the third stage oxytocin helps to obliterate the spiral arteries in the uterine wall and initiate lactation. Oxytocin can be synthesized and is often used to start contraction of the uterus if labour is delayed.

Hydatidiform mole Complication of an abortion or pregnancy due to continued growth of the chorion.

Jaundice Yellow discoloration that occurs when the bilirubin in the circulation rises above a certain level.

Miscarriage *See* Abortion.

Multiple pregnancy Homozygous—Identical cells, owing to mitotic division of the fertilized ovum forming separate units. Heterozygous—Fertilization of two or more ova at once.

Placenta Previa—Placenta is situated in front of the presenting part and may cover the internal opening of the cervix. Retained—Complication of the third stage of labour in which the placenta does not separate normally from the uterus.

Varicose veins Distension of the veins in the legs due to pressure of the presenting part on the venous return in the pelvis.

Vomiting Early-morning vomiting is a common feature in the first trimester of pregnancy.

Lactation

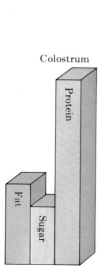
Colostrum

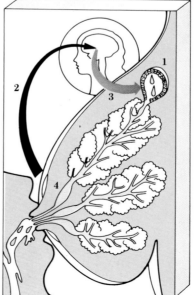

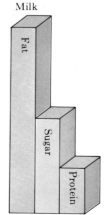
Milk

Growth 1

Definitions

Infancy

"Grasp" reflex Instinctive holding of an object in a baby's palm.

"Rooting" reflex Instinctive turning of a baby's head towards something that touches its cheek.

"Step" reflex Instinctive raising of a baby's foot when both are on a surface.

"Suck" reflex Instinctive sucking when something is placed in the mouth.

Teeth

Canine One of four teeth with a single point projecting above the level of the adjacent teeth.

Cementum Adherent connective tissue holding the ligament of the tooth to the jaw bone.

Crown Enamel-coated portion of the tooth above the level of the gum.

Deciduous "Milk" or temporary teeth.

Dentine Hard bone-like structure that forms the substance of the tooth.

Dentition Number and arrangement of teeth.

Enamel Coating of the crown of a tooth.

Incisor Four cutting teeth in the front of each jaw.

Molar Three permanent grinding teeth on each side of both jaws.

Neck Area of the tooth between the crown and root.

Premolar Two deciduous and permanent grinding teeth on each side of both jaws.

Pulp Soft centre of a tooth containing blood vessels, lymphatics and nerves.

Root One or more extensions of the tooth into the bone.

Locomotion and motor control

Newborn, fetal position

One month, legs stretching

Six weeks, lifts head

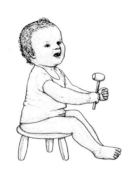

Six months, sitting

Ten months, crawling

Fifteen months, first steps

Infancy

The new-born baby lies with its knees drawn up. It reflexly grasps any object that touches the palm and when held upright automatically steps as the feet touch something. It roots and sucks the nipple automatically. These reflexes disappear within a few weeks. At a month the legs are straighter and by six weeks the head is lifted. The baby sleeps more often than not, but gradually the eyes move to focus on objects and at about six weeks smiling begins. By six months the birth weight is doubled and the child can sit unaided. At eight months the preliminary gurglings of speech are heard and the thumb can be used. At about ten months crawling starts; the birth weight is trebled. The first step may be taken at a year old. The first words may be spoken during the next two or three months.

Milk teeth (20)

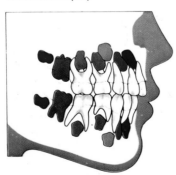

Teeth

Deciduous teeth start to appear at about six months; by the age of three the child has 20 teeth (*above left*)—8 incisors, 4 canines, 8 premolars.

Between the ages of six and twelve years the deciduous teeth are shed and replaced by the permanent dentition (*above centre*). A further six teeth, the molars, will appear in each jaw, and by the age of 25 a total of 32 teeth will be present.

The tooth consists of three parts (*above right*)—the crown, neck and root. The crown consists of dense mineral—enamel—surrounding the

Permanent teeth (32)

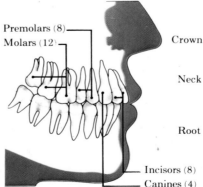

Premolars (8)
Molars (12)
Crown
Neck
Root
Incisors (8)
Canines (4)

Structure of a tooth

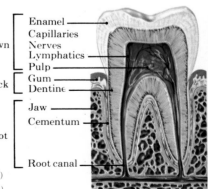

Enamel
Capillaries
Nerves
Lymphatics
Pulp
Gum
Dentine
Jaw
Cementum
Root canal

hard dentine, which has a soft centre —the pulp; the pulp is filled with blood vessels, lymphatics and the nerve, which reach it through the root canal. The neck adheres to the gum, and the root penetrates the bone, where it is held in place by a ligament and cementum.

Stages of development from birth to puberty

	Body	Senses	Mind	Speech	Social	Emotional	Moral
At birth	Almost immobile, but with many primitive reflexes e.g. sucking, grasping and rooting.	Even at 3 to 4 days old a baby is able to follow sound from side to side and can tell a buzzer from a bell.	Rapidly learns to turn head to left or right to find food in response to a buzzer or a bell.	Crying or gurgling are the only sounds.	Can be divided into active, moderately active and quiet babies.	Feelings are almost undifferentiated. Babies are awake or asleep, active or inactive.	Newborn is amoral, having no values, no attitudes and no beliefs for the first 6 or 7 months.
2 months	Can focus eyes and co-ordinate stare. Can begin to lift chin when lying on stomach.	Loud noises may be painful. Baby smiles at a picture of a simplified face made up of dots and angles.	Forgets existence of object if blocked for more than 15 seconds.	The first 3 months is usually the period of maximum crying.	Has often established a distinct pattern of crying and sociability. Starts smiling at his mother at 4 to 6 weeks.	Beginnings of the differentiation of feelings into positive and negative.	—
2-4 months	Able to raise head when lying on stomach and can turn it towards direction of speaker.	Shows signs of recognizing things. Can now appreciate 3-D.	Can tell mother from others but no idea that he has one mother only. Optical illusion of 3 delights him.	At beginning of period cooing starts. Responds to human voice and smiles when talked to. Sometimes chuckles.	Plays with objects (e.g. rattles) if placed in his hand. Beginnings of differential response to people around him.	At beginning of period first clearly defined signs of pleasure (smiling, chuckling, etc.) and negative emotions.	—
4-6 months	Opens hand for contact with an object and grasps things using palm and fingers but not the thumb.	Brings things to mouth in order to explore them. Smiles at moving picture of a face.	Child now distressed when shown optical illusion of 3 mothers.	Cooing becomes more vowel-like and consonants also appear.	Beginnings of real differentiation between adults at end of period.	Pleasure responses to others become selective. Fear and anger begin to emerge separately. Signs of self-satisfaction.	—
6-8 months	Can sit up without support for a short time, and stand with help. Can grasp using forefinger and thumb.	Attention begins to focus on expression rather than features of a face drawing.	Realization that hidden or out-of-sight objects continue to exist even when not seen.	Cooing begins to turn into babbling. Most common sounds are ma, mu, dar and di.	Beginnings of serious play but rarely with other children.	Fear of the unknown becomes developed. Attachment to known adults, and relationships becomes focused.	
8-10 months	Can crawl by the end of the period and pull himself upright to stand. Can hold feeding bottle on his own.	Depth perception is further developed. Child will avoid a visible drop from a surface.	Greater understanding of "object permanence". Child will search for objects in many different locations.	Repetitions of heard sounds become frequent but the results are often inaccurate.	Prefers to play with other people rather than by himself. Starts to imitate others.	Shows timidity towards strangers who are too familiar. Smiles at own image in mirror.	May show signs of withdrawal when admonished for bad conduct.
10-12 months	Can walk with help. Begins to stop putting objects in mouth. Bowel action starts to become more regular.	No longer likely to be upset by sudden, unexpected disappearance of an object.	With his first words begins to create and use symbols towards end of period.	First words are often nouns which serve multiple uses, i.e. "dog" may be used to describe any four-legged animal.	May show dramatic decrease in excitability and in attention and response when mother is present.	Differences between boys and girls seen in way they assert themselves and in readiness to touch others.	Understanding of parents' commands remains fairly primitive and tied to immediate situations.
12-18 months	Unaided walking at about 13 months. Helps to dress himself. Feeds himself. Creeps backwards down stairs.	Rapid increase in ability to use characteristic features of objects for identification purposes.	Up to 7 years increasingly links image of object with thought about it.	Repertoire of 3-50 words. Speech is largely telegraphic, e.g. "all-gone-food".	At end of period fighting and jealousy over toys emerges. Pleased with particular events or people.	Fear of strange objects, people and noises declines. At end of period shows jealousy over newly arrived babies.	Behaviour up to 6 or 7 years is shaped by threat of punishment or satisfaction of needs.
2 years	Runs but falls often. Climbs stairs holding rail. Girls reach half adult height after boys, just before 2 years old.	Can recognize displacement even though he has not seen object moved.	Thinks everyone has same view of object as he has, i.e. if covers eyes and cannot see you, you cannot see him.	Can string two words together (noun and qualifier). Grammatical use of articles, plurals, etc.	Solitary play declines. Parallel play with little contact between children in one room becomes more common.	At end of period fear of imaginary creatures and dark begins. More sensitive to ridicule.	Child's comprehension of parents' moral code consists of simple good/bad distinction.
2-3 years	Control over sphincter muscles allows toilet training. Bed wetting increasingly rare. Self-feeding improved.	Shows ability to distinguish between primary body parts and features; developing skill at drawing them.	Progress in ability to classify many sorts of objects according to one or more shared characteristics.	Over period able to put 3 or 4 words together. Structuring of language begins, use of tenses, plurals, order, etc.	By 30 months will be helping with domestic chores. Co-operation between children playing together increases.	Boys express more physical assertiveness and anger than girls. Attempts by both sexes at more independence.	Child likely to show substantial guilt-like reactions, though still situation-specific.
3-5 years	At end of period girls are slightly ahead of boys in skeletal development. The brain is 75% of its adult weight.	Often confuses the letters b and d, p and q. Ignores straightness, so confuses D and O.	Realizes apparent change (e.g. water/ice) is not real change, but is unlikely to understand why this is so.	Sentences become longer and more complex. Increasing grasp of basic grammar, like active-passive.	Peers start to influence child's behaviour. Identification with same-sex parent is at its strongest.	Increased satisfaction at accomplishing a self-set task. Rising sensitivity to feelings and responses of others.	Still highly egocentric but beginning to show marked guilt and self-regulation.
5-7 years	At 6 years old child's brain is 90% of its mature adult weight. Nerve fibres, etc., are almost complete.	Focuses on significant features of object or event. Increasingly able to discriminate between letters of the alphabet.	Child can use concepts and rules, but deals almost exclusively, in the beginning at least, with here and now.	Masters irregular endings (i.e. mice rather than mouses). Vocabulary becomes more complex and school related.	Turns to peers rather than adults. Play teaches social roles and individual limits.	Child actively engaged in testing self-image.	Child's judgement of "right" changes. Respect for, rather than fear of, law.
7-11 years	Rise in strength, speed and co-ordination. Skills now include those that need instruction and practice, e.g. skiing.		Can now conserve and see parts of the whole. Is able to arrange objects according to size, weight, etc.	By end of period has almost complete mastery over complex grammatical rules.	Social interactions strongest with members of the same sex; minimal social interaction between sexes.	Becomes less physical and more verbal in relationships with peers.	Child does things either to win approval or because the law says so.
11-13 years	Girls put on sudden growth spurt and by end of period they average 2½ years ahead of boys. Menstruation begins.	Increasingly is able to perceive non-obvious relationships among apparently unrelated events and objects.	Able to think around a subject, testing many solutions in his mind. Use of abstract rules to solve problems.	Ever widening vocabulary of more technical and specialized words. Greater range of terms to describe phenomena.	The sharing of possessions, feelings and plans is still primarily with same-sexed peers.	Alternating periods of withdrawal and gregariousness. Sexual feelings start to emerge.	Actions are guided according to universal ethical principles.

Growth 2

Puberty

Puberty is the moment when sexual maturation has reached the point where reproduction is possible. A period of development around puberty takes about two years and will usually start earlier in girls than in boys.

In girls, between 10 and 16, there are changes in the subcutaneous fat —hips and shoulders become more rounded, the breasts develop and the pubic and axillary hair grows. The first menstruation—menarche —is a dramatic event establishing a definite point of maturity.

In a boy, between 12 and 17, the shoulders broaden, the muscles strengthen and the genitalia develop and darken to become covered with pubic hair. The larynx lengthens and the voice "breaks" to become deeper. Spontaneous erections occur; the nocturnal emissions of sperm are less dramatically the sign of sexual maturity than the menarche.

Adolescence

Adolescence is the time of greatest physical and emotional change, which can be disturbing not only to the individual concerned but also to the family. It is a time when social, intellectual and sexual interests expand and broaden. The young person, no longer a child and already more or less physiologically mature, must gradually loosen the ties associated with the home and parents. This involves a questioning of parental wisdom and morality, which is often replaced by less desirable leadership and a tendency to truancy and even crime.

The intense emotions of adolescence are often released as an obsessive or hysterical hero worship, particularly of such heroes as sportsmen and pop stars. Screaming at pop concerts, for example, is one way the adolescent can release pent-up emotions. Such periods of great enthusiasm may be followed by ones of almost depressive lethargy.

The stresses of adolescence and the feeling of independence may lead to experimentation with drugs as an escape from the conflicts of a demanding adult world. This turbulence may lead to depression.

The young adult

As the physical and psychological turmoil of adolescence settles young adults are at their prime. There is a better relationship with family and society and a greater awareness and control of emotions and sexual desire. The first tentative adolescent meetings with the opposite sex have passed and mature relationships develop. Work may give a feeling of achievement and independence—a confidence previously lacking.

Middle age

Usually the most stable period of life starts around the age of 30. It is usually a time of relative financial and emotional security built around the family and growing children, in those who have married, or friends and work in those who remain single. Sexuality is less urgent but often more satisfactory.

In the woman the menopause— the final cessation of menstruation— may be followed by a time of hormonal disturbance leading to hot flushes, tensions and sometimes depression. If this coincides with the storm of her children's adolescence and her husband's demands at work, she may feel her life is a failure; in fact, it is a time of achievement.

Although physical health usually remains good there is a tendency to gain weight, accompanied by a slight, but definite loss of athletic skills. In the later years there is an increase in the death rate, particularly from heart disease and cancer of the lungs.

Old age

In Western society old age seems to start arbitrarily on retirement. This is a psychological shock, as many still feel fit, active and well and find difficulty in accustoming themselves to days of idleness and boredom, often with increasing problems of financial anxiety. The minor stiffness of early arthritis, the slight loss of visual acuity and minimal deafness may produce a slower reaction time, so that falls are more common.

The adrenal glands continue to secrete sex hormones and the slight androgenic effect may produce hair growth and slight baldness in women, while in men there may be the estrogen effect of female fat distribution and less beard growth. Cancer and other illnesses are common and often fatal.

Senescence

This is the period of real ageing and may begin seriously before or long after the arbitrary point of retirement. The skin loses its elasticity and gains wrinkles; the hair loses its colour, becoming white as it fills with microscopic bubbles of air; the senses of vision, hearing and taste deteriorate; the joints become rough and arthritic, while the muscles weaken. The bones in the skeleton soften as the structure deteriorates and calcium is removed, causing the bent spine of the elderly.

The nervous system is involved with loss of nerve cells, leading to failure of memory, first for recent events and later a gradual dementia may occur. The cerebellum is affected causing a loss of co-ordination, tremor of the hands and hesitation in speech. The heat centre in the thalamus is less responsive and a cold environment may cause a drop in body temperature—hypothermia— that can lead to coma and death. Deterioration in the heart and lungs may produce irregularities of cardiac rhythm and often breathlessness.

Dietary problems are not only financial but also practical, involving preparation of the correct kinds of food when arthritic, absentminded and breathless. Chewing is difficult and failure to eat roughage will lead to constipation and abdominal discomfort.

In some people these ageing processes take place faster in one system than another. This can be seen with dementia occurring while the body is fit or osteo-arthritis crippling an otherwise healthy woman.

Illness and disease may bring death before the possibility of social problems, such as the paranoid, incontinent and arthritic grandparent who fails to eat properly and refuses help. These social, economic and family pressures of senescence are enormous and need help from visiting nurses, social workers and the family doctor unless admission to hospital is the only way of giving sufficient basic nursing care.

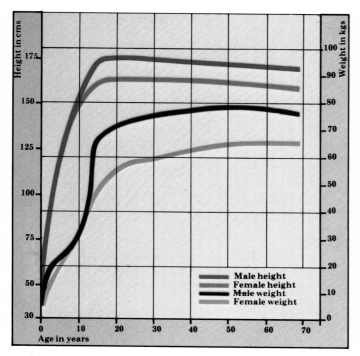

Weight and height The graph shows a steady increase in males and females until the early teens, when there is a short growth spurt reaching a maximum. The height loss with ageing ·is caused by loss of spinal bone density.

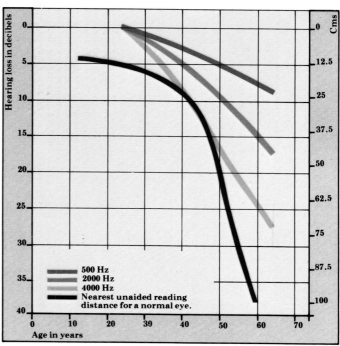

Hearing and vision As one ages the nearest point on which the unaided eye can focus dramatically recedes around the age of 40. The ability to hear also declines with the highest pitches deteriorating faster than lower frequencies.

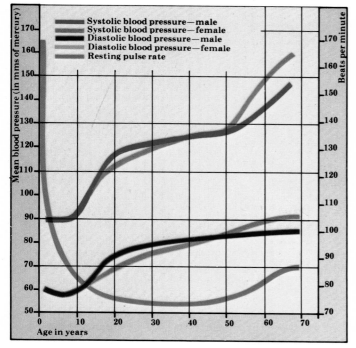

Pulse and blood pressure The rapid pulse rate of infancy falls to a low rate during adulthood and is accompanied by a rise in blood pressure that is greater in systolic than diastolic levels, caused by arterial hardening.

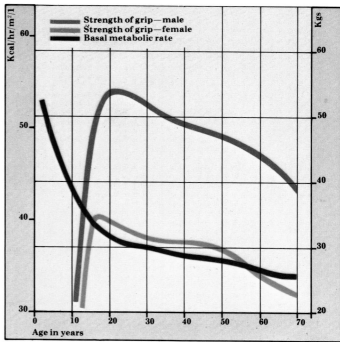

Physical strength and basal metabolic rate The development of greater strength in males is followed by steady deterioration with age. The metabolic rate declines rapidly at first and then slows later in life as cell metabolism declines.

Metabolism and Nutrition 1

Definitions

Adrenaline Hormone produced by the adrenal medulla.

Amino-acid One of the basic units from which proteins are made. Eight essential amino-acids cannot be synthesized by the body: Iso-leucine, leucine, lysine, methionine, phenylalanine, threonine, tryptophan and valine. All the others can be synthesized.

Ammonia Toxic nitrogen-containing substance produced by the liver in the breakdown of proteins which is transformed into urea for excretion in the urine.

Anti-diuretic hormone Hormone produced by the posterior lobe of the pituitary gland which acts on the kidney to increase the reabsorption of water.

Appetite centre Area in the hypothalamus of the brain that controls the balance between the body's requirements for food and its use of energy.

Basal metabolic rate Speed with which energy is used when an individual has been at rest, physically and mentally, for twelve hours. It is measured in the hourly calorie requirements for each square metre of body surface.

Calorie Nutritionally this is taken to be the kilo-calorie. A calorie is the amount of heat required to raise 1 gramme of water through 1 degree centigrade. A kilo-calorie is 1,000 times this.

Carbohydrate Basic structure of starch or sugar.

Catabolism Breakdown of complex body substances into simpler compounds.

Cholesterol Substance produced by the breakdown of fats.

Cingulate gyrus Part of the limbic system, associated with memory.

Food requirements

The body has certain basic daily food requirements to maintain health, replace old tissues, and for growth and energy. A reserve must be kept for unexpected stresses. The main constituents of a diet are proteins, carbohydrates, fats, vitamins, minerals, roughage and water.

Protein is the nitrogen-containing body-building material found in most tissue, enzymes and muscles. It is digested in the form of 20 different amino-acids, which are required to make all the proteins in the body. The liver can synthesize most of these, but eight, the essential amino-acids, have to be obtained from the diet. Complete proteins are those that contain these essential amino-acids, such as meat, fish and eggs; incomplete proteins, such as beans, peas and pulses, only contain some. A diet of several incomplete proteins will supply sufficient amino-acids. An active individual needs one gramme of protein for each kilogramme of weight every day.

Protein is lost in the body as urea, creatinine, uric acid and ammonia; it is also lost as hair and dead cells from the skin and intestine. It can be used for energy or changed into fat.

Carbohydrates form part of the basic body structure as glycoprotein. They are stored as glycogen in the liver and used for energy, being metabolized into carbon dioxide and water. This metabolism is controlled by insulin and glucagon and also affected by the hormones from the adrenal and thyroid glands. Carbohydrates are found in food as starch and the sweet sugar, sucrose. Excess amounts are stored as fat.

Fats are found in the cell fabric as lipo-proteins and carried in the circulation as phospholipids. They are found in many foods, animal fats having a high content of cholesterol and saturated fatty acids, unlike vegetable oils. Fats are stored throughout the body in fat cells and are used for energy, producing water, carbon dioxide and, if too much is used, acetic acid and ketones. They are an important store of reserve energy and heat.

Vitamins are substances required in minute amounts for the normal working of the body. Fat-soluble vitamins are A, D, E and K and water-soluble in the vitamin B group; vitamin C cannot be stored. Minerals include calcium, phosphorus, sodium, potassium and carbon.

Small quantities of iron, copper, cobalt, iodine, manganese, magnesium and fluorine are needed in certain enzyme reactions, blood formation or the teeth.

Roughage is needed, in the form of cellulose and vegetable fibre, to supply bulk to the feces and ensure normal defecation.

At least 2,500 millilitres of water are needed each day for normal health. Water is obtained and excreted in the following way:

Intake		Excretion	
Drink	1,500ml	Urine	1,500ml
Food	750ml	Respi-	
Metabo-		ration	400ml
lism	250ml	Sweat	500ml
		Feces	100ml
Total	**2,500ml**	**Total**	**2,500ml**

Energy requirements

A normal, healthy, balanced diet provides all energy requirements. A nursing sister requires much more energy than a sedentary office worker. Energy is measured in calories, or kilojoules, and is supplied from the food. One gramme of protein or carbohydrate supplies four calories, while one gramme of fat supplies nine calories. The specific dynamic action (SDA) of food is the amount of energy required to release the potential energy in the diet. This is the energy taken to digest, absorb and change substances before they can be used; for example, to obtain 100 calories of energy from 100 calories worth of fat or carbohydrate, 105 calories worth of food has to be eaten. Proteins have a higher SDA of 125, hence a high protein, low carbohydrate diet will release fewer calories of useful energy than an equivalent weight of carbohydrate.

The basal metabolic rate (BMR) is the speed with which energy is used when the individual has been resting physically and mentally for at least 12 hours after the last meal. It is measured in the hourly calorie requirements for each square metre of body surface. In a healthy adult woman it is about 38. The BMR is increased by six main factors:

Age—highest BMR in a very young child and lowest in old age.

Pregnancy—increased BMR in the mother to account for the baby's BMR.

Sex—greater BMR in males.

Food—the SDA of diet increases the BMR.

Temperature—fever increases BMR by 10 per cent for each 1°C rise. External chilling will increase the BMR to produce the extra heat required to maintain the body temperature.

Hormones and chemicals—increased thyroid production, excess caffeine, or amphetamines stimulate cell activity.

The BMR is decreased by four main factors: Sleep; starvation; low thyroid activity; and prolonged, excessive chilling.

The circadian biological rhythms of the body cause regular fluctuations in the BMR. The appetite centre maintains a reasonably constant body weight by balancing the dietary intake with the metabolism.

The major food categories

Carbohydrates	Proteins	Fats	Roughage	Liquids	Vitamins and minerals
Sugary Honey, bananas, grapes, beets. *Starchy* Wheat, rice, potatoes, yams, cassava.	*Complete* Meat, fish, cheese, milk, yoghurt, eggs. *Incomplete* Lentils, beans, wheat, bread, nuts, peas, potatoes.	Milk, cream, cheese, butter, margarine, cooking oil, nuts, soya beans, fatty fish, bacon, eggs, milk chocolate.	Skins, peeling and pulp of fruit and vegetables, husks of grain.	*Drinks* Juices, milk, water. *Contained in food* Green vegetables, vegetable fruits (e.g. tomatoes).	*See* accompanying charts. Varied diet gives enough for average daily adult requirements.

The major vitamins and minerals

Vitamins or minerals	Good sources	Daily adult requirement	Function	Notes of interest
Vitamin A (retinol)	Liver, fish liver oil, eggs, butter, leaf vegetables, carrots, apricots.	750μg (e.g. 2 carrots or 2 eggs)	Night vision, healthy skin and mucous membranes.	Toxic in large doses as stored in the liver. Deficiency causes night blindness and dry skin.
Vitamin B₁ (thiamin)	Meat, cereals (particularly wheat germ), vegetables, yeast.	1.5mg (e.g. 1 cup wheat germ)	General metabolic processes, normal appetite, digestion and functioning of the nerves.	Deficiency causes beriberi.
Vitamin B₂ (riboflavin)	Liver, kidney, milk, eggs, yeast.	2mg (e.g. 4oz liver)	General metabolic processes, healthy skin and mucous membranes.	Ultraviolet light destroys it in the milk bottle. Deficiency is rare, but it causes cheiliosis.
Nicotinic acid or nicotinamide (niacin)	Meat, fish, cereals, whole-wheat flour, yeast and bran.	20mg (e.g. 2oz brewers' yeast)	As for vitamin B₂.	The body can make it from the amino-acid tryptophan. Deficiency causes pellagra.
Folic acid	Liver, fish, cereals, raw greens, vegetables, yeast.	200μg (e.g. 6oz liver)	Formation of new cells such as red blood cells and metabolism of nerve cell.	Deficiency causes anemia, especially in pregnancy, and peripheral neuritis.
Vitamin B₆ (pyridoxine)	Meat, fish, nuts, cereals, yeast.	2mg (e.g. 8oz chicken)	Metabolism of amino-acids and haemoglobin, healthy skin and mucous membranes.	Deficiency causes peripheral neuritis.
Vitamin B₁₂ (cyanocobalamin)	Liver, fish, eggs, cheese.	5μg (e.g. 4oz fatty fish)	Metabolism of nerve cells and the production of red blood cells.	Deficiency causes pernicious anemia and peripheral neuritis.
Vitamin C (ascorbic acid)	Green vegetables, potatoes, citrus fruits and green peppers.	30mg (e.g. 1 orange)	Tissue growth and repair and the production of steroid hormones.	Much is lost in cooking. Deficiency causes scurvy.
Vitamin D	Fatty fish, cod liver oil, eggs, fortified margarine.	2.5μg (e.g. 2oz fatty fish)	Correct absorption of calcium and phosphorus and their metabolism.	Deficiency causes rickets in children and osteomalacia in adults.
Vitamin E (tocopherol)	Eggs, cereals, margarine, vegetable oils.	Not known	General metabolic processes and the metabolism of nucleic acids.	May improve fertility.
Vitamin K (menadione)	Cereals, green vegetables, pulses.	Not known	Clotting of the blood.	Some is made by intestinal bacteria. Stops bleeding if anti-coagulant drugs used.
Pantothenic acid, choline, biotin	Nearly all foods.	Low	Fat metabolism.	Intestinal bacteria may manufacture most of these.
Calcium	Milk, cheese, bread, sardines, green vegetables, watercress.	500mg (e.g. 2 cups milk)	Healthy bones and teeth, muscle contraction, blood clotting, nerve transmission and activity of some enzymes.	Need is double during pregnancy. Deficiency causes rickets in children and osteomalacia in adults.
Iron	Meat, poultry, eggs, fruit, leaf vegetables, yeast.	10mg (e.g. 4oz liver)	Production of hemoglobin and myoglobin.	Need is higher during menstruation and pregnancy. Deficiency causes anemia.
Phosphorus	Meat, fish, cheese, eggs, nuts, pulses, rice, bread, yeast.	Sufficient in normal diet	Healthy bones and teeth. Formation of nucleic acids, ATP and cell membranes.	Second commonest mineral in the body.
Sodium and Chloride	Table salt, seafoods, eggs, milk, most foods except fruit.	Sufficient in normal diet	Part of body fluids.	Need is higher with excessive sweating and vomiting.
Potassium	Meat, milk, cereals, vegetables, fruit.	Sufficient in normal diet	Part of body fluids. Nerve and muscle function.	Need is higher after severe diarrhea.
Fluorine	Drinking water, tea, coffee, rice, vegetables.	Varies according to area	Healthy bones and teeth.	Deficiency increases dental decay.
Iodine	Sea foods, iodized salt, water and vegetables in non-goitrous regions.	Varies according to area	Production of thyroid hormones.	Deficiency causes goitre, cretinism and myxedema.
Zinc	Meat (liver), shellfish, milk, wheat, bran.	Sufficient in normal diet	Enzymes.	Need may be higher during lactation.
Copper	Meat, poultry, shellfish, nuts, wholegrains.	Sufficient in normal diet	Enzymes. Production of blood.	Deficiency very rarely causes anemia in babies.
Cobalt	Meat, poultry, shellfish, vegetables.	Sufficient in normal diet	Part of vitamin B₁₂.	
Magnesium	Meat, milk, nuts, cereals, vegetables.	Sufficient in normal diet	Nerve and muscle function. Enzymes.	Deficiency may occur after prolonged diarrhea.
Manganese	Wholegrains, nuts, tea.	Sufficient in normal diet	Manufacture of fats. Release of energy in cells.	

Circadian Daily rhythm.

Cortisol Hormone produced by the adrenal cortex.

Creatinine Nitrogen-containing waste product of metabolism.

Fatty acid Basic constituent of fat.

Gastrin Hormone produced by the stomach to maintain the gastric secretions.

Glucagon Hormone produced by the pancreas that increases the level of glucose in the blood.

Glycogen Form of starch.

Glycoprotein Combination of glycogen and protein that helps in the formation of cells.

Glycerol Part of the basic constituent of fats.

Hemoglobin Complex protein that gives the colour to red cells and is essential for the carriage of oxygen. It contains the element iron.

Hippocampus Part of the limbic system associated with memory.

Hypothalamus Lying below the thalamus in the brain, it is concerned with the nervous control of the pituitary gland in its relationship with the vital centres.

Insulin Hormone produced by the pancreas that reduces the level of glucose in the blood.

Ketosis Increase in ketones in the blood, during excessive fat metabolism.

Kilo-joule Modern measurement of energy based on the Système Internationale d'Unités: 4·2 kilo-joules 1 calorie.

Limbic system Pair of structures lying above the thalamus which are concerned with memory and instinctive emotions.

Metabolism and Nutrition 2

Lipo-protein
Combination of fat
and protein that is
used in the construc-
tion of cells.

Mineral Non-
organic substance,
e.g. iron and
phosphorus.

Nor-adrenaline
Hormone secreted by
the medulla of the
adrenal gland.

Olfactory lobe
Area of the frontal
lobe concerned with
smell.

Phospholipid
Form of fatty acid
containing
phosphorus.

Protein One of a
variety of nitrogen-
containing com-
pounds composed of
amino-acids. First-
class proteins are
those which contain
all the essential
amino-acids. Second-
class proteins are
those which contain
many, but not all, the
essential amino-acids.

Diet and the whole body

The healthy body is stimulated by
the thought, sight and smell of food.
The brain co-ordinates these stimuli.
The limbic system and the hippocampus and cingulate gyrus help to
activate the appetite centre in the
hypothalamus. The olfactory lobes
and visual cortex relay messages
through the thalamus and mid-brain
to initiate salivation and gastric
secretion via the autonomic nervous
system. At the same time a falling
blood sugar and the biological
rhythms of the body may coincide
and increase the sensation of hunger.

These stimuli are reinforced by
eating, which activates the taste
buds in the tongue and thus increases
further salivation. The swallowing of
food into the stomach sets in action
gastric secretion and gastric hormonal release—gastrin—to maintain
gastric and initiate duodenal secretions. This, in turn, produces further
hormones to activate pancreatic and
bile production.

In the graph (below) nine periods
in the life of a man or woman have
been selected. Each block represents
an essential nutrient in a healthy
diet. The intake of a nutrient can be
compared as a percentage of the
maximum ever required and the
quantity by the thickness of the
appropriate block.

The extraction of these nutritional
elements occurs after the enzyme
digestion. They are either used immediately to replace and repair worn-
out and damaged tissue or transformed into energy and dissipated as
heat. Some are stored in the liver or
transformed into fat and transferred
to the fat depots throughout the
body. The carbon dioxide produced
is exhaled by the lungs; the water is
excreted by the kidneys and the
amino-acids either broken down into
urea, uric acid, creatinine and ammonia, or reformed into new protein
by the liver. Cholesterol is excreted
in the bile.

All this activity is dependent on

small amounts of some of the essential nutrients to help maintain the
intra-cellular enzymes in healthy
metabolic order. The connective
tissue starts to disintegrate without
vitamin C—ascorbic acid; the
hemoglobin fails to form without
adequate supplies of iron and calcium absorption; and bone production is impaired without vitamin D.

The metabolism of every cell depends upon receiving the right substances via the liver and the circulation. Hormones balance this activity,
many of them under the overall
control of the anterior lobe of the
pituitary gland. As the diet is
digested and assimilated into the
healthy body the activity of the
reticular system in the brain stem is
reduced and a feeling of contentment
and drowsiness begins to cloud the
consciousness. The heat produced by
the metabolic energy of digestion
and the specific dynamic action of
the various foods adds to the warm
contentment of a satisfactory meal.

Recommended daily intake of major nutrients

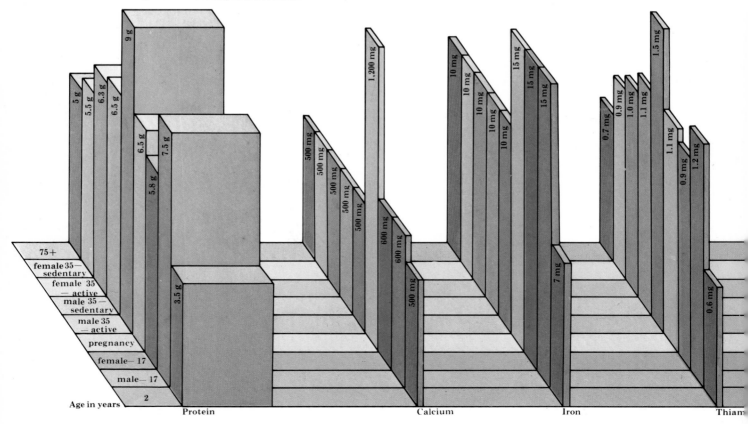

92

Energy requirements

Energy requirements

The body's need for energy from the diet varies not only with activity, sex, disease and climate, but also with age. The size of the figures in the diagram (*above*) show the comparative energy requirements from birth to adulthood. Up to two years old, the rapidly growing child needs more than anyone in proportion to size; by old age, when metabolism is slowest, the need is far less.

Vitamin A
(Retinol equivalents)

750μg · 750μg · 750μg · 750μg · 1,200μg · 750μg · 750μg · 300μg

Vitamin C

30 mg · 30 mg · 30 mg · 30 mg · 60 mg · 30 mg · 30 mg · 20 mg

Vitamin D

2.5μg · 2.5μg · 2.5μg · 2.5μg · 10μg · 10μg · 2.5μg · 2.5μg

Calories

1,900 · 2,200 · 2,500 · 2,700 · 3,600 · 2,600 · 2,300 · 3,000 · 1,400

Septum pellucidum Part of the limbic system concerned with pleasurable emotions.

Sleep centre Situated behind the thalamus of the brain, it suppresses the activity of the reticular formation.

Specific dynamic action Amount of energy required to release the potential energy in food. More energy is used in metabolism of protein than either fat or carbohydrate, which means that protein gives less useful energy to the body.

Starch Form of carbohydrate.

Sucrose Simple sugar.

Thalamus Mass of grey matter at the base of the brain concerned with sensory nervous transmission.

Thyroid gland Endocrine gland in the neck which produces hormones to stimulate the body's metabolism.

Urea Simple nitrogen-containing substance that is a body metabolite.

Uric acid Simple nitrogen-containing substance that is a body metabolite; excess amounts will cause gout.

Visual cortex Situated at the back of the cerebral hemisphere, it is the area of grey matter concerned with vision.

Vitamin One of a series of substances that are found in the normal diet and are required in small amounts for normal metabolism.

Index

Index